CARBON NANOTUBES AND RELATED STRUCTURES

New Materials for the Twenty-first Century

This book covers all the most important areas of nanotube research, as well as discussing related structures such as carbon nanoparticles and 'inorganic fullerenes'.

Carbon nanotubes are molecular-scale carbon fibres with structures related to those of the fullerenes. Since their discovery in 1991, they have captured the imagination of physicists, chemists and materials scientists alike. Physicists have been attracted to them because of their extraordinary electronic properties, chemists because of their potential as 'nanotest-tubes', and materials scientists because of their amazing stiffness, strength and resilience. On a more speculative level, nanotechnologists have considered possible nanotube-based gears and bearings.

This is the first single-author book on carbon nanotubes. It will be of interest to chemists, physicists, materials scientists and engineers working on carbon materials and fullerenes from both academic and industrial backgrounds.

PETER HARRIS was brought up in Gloucestershire and read chemistry at Birmingham University. He went on to study for a doctorate at Oxford University, where his project involved transmission electron microscopy of catalytic materials. Since that time his research has focused on the application of various forms of microscopy to problems in solid-state chemistry and materials science. He has carried out post-doctoral work at both Cambridge and Oxford, and currently works in the Chemistry Department at Reading University, where he is responsible for electron microscopy. In addition to his work on carbon nanotubes and nanoparticles, he is involved in a wide range of projects for departments across the University. He has published over 40 scientific papers, and regularly reviews books for materials and microscopy journals. He lives in Twyford, outside Reading, with his wife and two daughters.

CARBON NANOTUBES
AND RELATED STRUCTURES

New Materials for the Twenty-first Century

Peter J. F. Harris

Department of Chemistry, University of Reading

CAMBRIDGE
UNIVERSITY PRESS

PUBLISHED BY THE PRESS SYNDICATE OF THE UNIVERSITY OF CAMBRIDGE
The Pitt Building, Trumpington Street, Cambridge, United Kingdom

CAMBRIDGE UNIVERSITY PRESS
The Edinburgh Building, Cambridge CB2 2RU, UK www.cup.cam.ac.uk
40 West 20th Street, New York, NY 10011-4211, USA www.cup.org
10 Stamford Road, Oakleigh, Melbourne 3166, Australia
Ruiz de Alarcón 13, 28014 Madrid, Spain

First published 1999

Printed in the United Kingdom at the University Press, Cambridge

Typeset in Times 11/14pt [VN]

A catalogue record for this book is available from the British Library

Library of Congress Cataloguing in Publication data
Harris, Peter J. F. (Peter John Frederich), 1957–
Carbon nanotubes and related structures: new materials for the
21st century/Peter J. F. Harris.
p.   cm.
Includes bibliographical references.
ISBN 0 521 55446 2 (hc.)
1. Carbon.   2. Nanostructure materials.   3. Tubes.   I. Title.
TA455.C3H37   1999
620.1′93–dc21   99-21391 CIP

ISBN 0 521 55446 2 hardback

To Elaine, Katy and Laura

Contents

Acknowledgements

Writing a book on carbon nanotubes has involved delving into such unfamiliar areas (for a chemist) as the mechanical properties of composite materials and the behaviour of arc plasmas. None of this would have been possible without the freely given assistance of colleagues from a wide range of disciplines, many of whom have also provided copies of images and preprints. The following list almost certainly fails to include all who have helped me, so I apologise in advance for any omissions. I also stress that any errors which remain in the book are my responsibility alone.

I wish to thank: Pulickel Ajayan, Hiroshi Ajiki, Severin Amelinckx, Don Bethune, Florian Banhart, Adrian Burden, Peter Buseck, Jean-Christophe Charlier, Nasreen Chopra, Daniel Colbert, Cees Dekker, Millie Dresselhaus, Thomas Ebbesen, Malcolm Green, Simon Hibble, John Hutchison, Sumio Iijima, George Jeronimidis, Radi Al Jishi, Philippe Lambin, Charles Lieber, Annick Loiseau, Amand Lucas, David Luzzi, Sara Majetich, Madhu Menon, Youichi Murakami, Eiji Osawa, Zhifeng Ren, Riichiro Saito, Yahachi Saito, Klaus Sattler, Jeremy Sloan, Reshef Tenne, Mauricio Terrones, Andreas Thess, David Tomanek, Edman Tsang, Daniel Ugarte and Boris Yakobson.

I also want to thank Simon Capelin of Cambridge University Press for his patience and encouragement, and Margaret Patterson for her meticulous copy-editing.

Finally, I thank my wife and daughters for their love and support, and my father for all the advice he has given me on this book and so many other subjects over the years.

Peter Harris
Twyford

1

Introduction

Take Carbon for example then
What shapely towers it constructs
A. M. Sullivan, *Atomic Architecture*

Carbon, in fact, is a singular element . . .
Primo Levi, *The Periodic Table*

The ability of carbon to bond with itself and with other atoms in endlessly varied combinations of chains and rings forms the basis for the sprawling scientific discipline that is modern organic chemistry. Yet until recently we knew for certain of just two types of *all-carbon* crystalline structure, the naturally occurring allotropes diamond and graphite. Despite the best efforts of some of the world's leading synthetic chemists, all attempts to prepare novel forms of molecular or polymeric carbon came to nothing: the elegant all-carbon structures proposed by Roald Hoffmann, Orville Chapman and others remained firmly in the realm of pure speculation. Ultimately, the breakthrough which revolutionised carbon science came not from synthetic organic chemistry but from experiments on clusters formed by the laser-vaporisation of graphite.

Harry Kroto, of the University of Sussex, and Richard Smalley, of Rice University, Houston, had different reasons for being interested in the synthesis of carbon clusters. Kroto had been fascinated since the early 1960s in the processes occurring on the surfaces of stars, and believed that experiments on the vaporisation of graphite might provide key insights into these processes. Smalley, on the other hand, had been working for several years on the synthesis of clusters using laser-vaporisation, concentrating chiefly on semiconductors such as silicon and gallium arsenide. But he was also interested in what might happen when one vaporises carbon. In August 1985, the two scientists came together at Rice and, with a group of colleagues and students,

1

began the now famous series of experiments on the vaporisation of graphite. They were immediately struck by a surprising result. In the distribution of gas-phase carbon clusters, detected by mass spectrometry, C_{60} was by far the dominant species. This dominance became even more marked under conditions which maximised the amount of time the clusters were 'annealed' in the helium. There was no immediately obvious explanation for this since there appeared to be nothing special about open structures containing 60 atoms. The eureka moment came when they realised that a *closed* cluster containing precisely 60 carbon atoms would have a structure of unique stability and symmetry, as shown in Fig. 1.1. Although they had no direct evidence to support this structure, subsequent work has proved them correct. The discovery of C_{60}, published in *Nature* in November 1985 (1.1), had an impact which extended way beyond the confines of academic chemical physics, and marked the beginning of a new era in carbon science (1.2–1.5).

At first, however, further progress was slow. The main reason was that the amount of C_{60} produced in the Kroto–Smalley experiments was minuscule: 'a puff in a helium wind'. If C_{60} were to become more than a laboratory curiosity, some way must be found to produce it in bulk. Eventually, this was achieved using a technique far simpler than that of Kroto and Smalley. Instead of a high-powered laser, Wolfgang Krätschmer of the Max Planck Institute at Heidelberg, Donald Huffman of the University of Arizona and their co-workers used a simple carbon arc to vaporise graphite, again in an atmosphere of helium, and collected the soot which settled on the walls of the vessel (1.6). Dispersing the soot in benzene produced a red solution which could be dried down to produce beautiful plate-like crystals of 'fullerite': 90% C_{60} and 10% C_{70}. Krätschmer and Huffman's work, published in *Nature* in 1990, showed that macroscopic amounts of solid C_{60} could be made using methods accessible to any laboratory, and it stimulated a deluge of research.

Carbon nanotubes, the primary subject of this book, are perhaps the most important fruits of this research. Discovered by the electron microscopist Sumio Iijima, of the NEC laboratories in Japan, in 1991, these 'molecular carbon fibres' consist of tiny cylinders of graphite, closed at each end with caps which contain precisely six pentagonal rings. We can illustrate their structure by considering the two 'archetypal' carbon nanotubes which can be formed by cutting a C_{60} molecule in half and placing a graphene cylinder between the two halves. Dividing C_{60} parallel to one of the three-fold axes results in the zig-zag nanotube shown in Fig. 1.2(a), while bisecting C_{60} along one of the five-fold axes produces the armchair nanotube shown in Fig. 1.2(b). The terms 'zig-zag' and 'armchair' refer to the arrangement of hexagons around the circumference. There is a third class of structure in which the hexagons are arranged helically

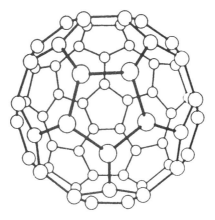

Fig. 1.1. C_{60}: buckminsterfullerene.

around the tube axis (see Chapter 3). Experimentally, the tubes are generally much less perfect than the idealised versions shown in Fig. 1.2, and may be either multilayered or single-layered.

Carbon nanotubes have captured the imagination of physicists, chemists and materials scientists alike. Physicists have been attracted to their extraordinary electronic properties, chemists to their potential as 'nanotest-tubes' and materials scientists to their amazing stiffness, strength and resilience. On a more speculative level, nanotechnologists have discussed possible nanotube-based gears and bearings. In this book, an attempt has been made to cover all of the most important areas of nanotube research, as well as discussing related structures such as carbon nanoparticles, carbon onions and 'inorganic fullerenes'. This opening chapter begins with a brief account of the discovery of carbon nanotubes and then describes some of the basic characteristics of arc-evaporation-synthesised nanotubes. The pre-1991 evidence for the existence of nanotubes is discussed, and some of the directions in which nanotube research is developing are summarised. Finally, the organisation of the book is outlined.

1.1 The discovery of fullerene-related carbon nanotubes

Iijima was fascinated by the Krätschmer–Huffman *Nature* paper, and decided to embark on a detailed TEM study of the soot produced by their technique. He had good reasons for believing that it might contain some interesting structures. Ten years earlier he had studied soot formed in a very similar arc-evaporation apparatus to the one used by Krätschmer and Huffman and found a variety of novel carbon architectures including tightly curved, closed

a

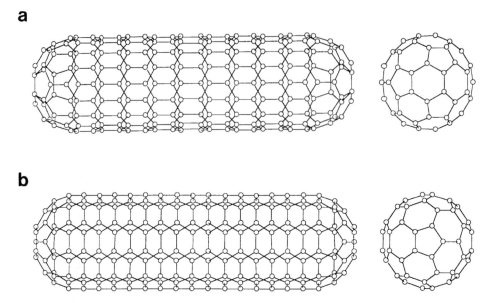

b

Fig. 1.2. Drawings of the two nanotubes which can be capped by one half of a C_{60} molecule (1.7). (a) Zig-zag (9,0) structure, (b) armchair (5,5) structure (see Chapter 3 for explanation of indices).

nanoparticles and extended tube-like structures (1.8, 1.9). Might such particles also be present in the K–H soot? Initial high resolution TEM studies were disappointing: the soot collected from the walls of the arc-evaporation vessel appeared almost completely amorphous, with little obvious long-range structure. Eventually, Iijima gave up sifting through the wall soot from the arc-evaporation vessel, and turned his attention to the hard, cylindrical deposit which formed on the graphite cathode after arc-evaporation. Here his efforts were finally rewarded. Instead of an amorphous mass, the cathodic soot contained a whole range of novel graphitic structures, the most striking of which were long hollow fibres, finer and more perfect than any previously seen. Iijima's beautiful images of carbon nanotubes, shown first at a meeting at Richmond, Virginia in October 1991, and published in *Nature* a month later (1.10), prompted fullerene scientists the world over to look again at the used graphite cathodes, previously discarded as junk.

1.2 Characteristics of multiwalled nanotubes

A typical sample of the nanotube-containing cathodic soot is shown at moderate magnification in Fig. 1.3(a). As can be seen, the nanotubes are accompanied by other material, including nanoparticles (hollow, fullerene-related struc-

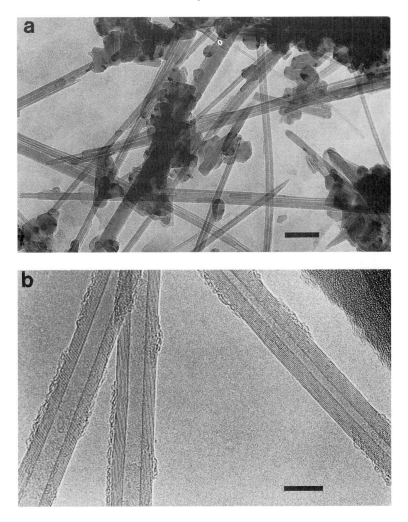

Fig. 1.3. (a) TEM image of nanotube-containing soot. Scale bar 100 nm. (b) Higher magnification image of individual tubes. Scale bar 10 nm.

tures) and some disordered carbon. The nanotubes range in length from a few tens of nanometres to several micrometres, and in outer diameter from about 2.5 nm to 30 nm. At high resolution the individual layers making up the concentric tubes can be imaged directly, as in Fig. 1.3(b). It is quite frequently observed that the central cavity of a nanotube is traversed by graphitic layers, effectively capping one or more of the inner tubes and reducing the total number of layers in the tube. An example is shown in Fig. 1.4, where a single layer forms a cap across the central tube, reducing the number of concentric layers from six to five.

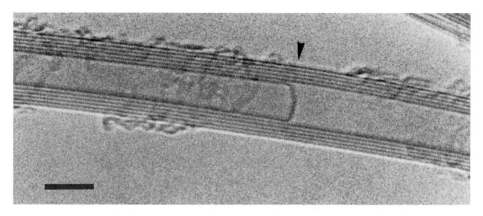

Fig. 1.4. High resolution image of multiwalled nanotube with 'internal cap'. Scale bar 5 nm.

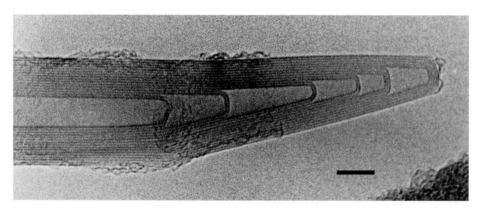

Fig. 1.5. Image of typical multiwalled nanotube cap. Scale bar 5 nm.

As mentioned above, virtually all of the tubes are closed at both ends with caps which contain pentagonal carbon rings. In practice, the caps are rarely hemispherical in shape, but can have a variety of morphologies; a typical example is shown in Fig. 1.5. More complex cap structures are often observed, owing to the presence of heptagonal as well as pentagonal carbon rings (1.11). Iijima has often illustrated the role played by pentagonal and heptagonal rings in nanotube caps by referring to the art of Japanese basket-work, of the kind shown in Fig. 1.6, where non-hexagonal rings play a similar topological role. Structures analogous to those of carbon nanotubes also occur among viruses (see Chapter 3), and, perhaps inevitably, among the architectural designs of Buckminster Fuller (Fig. 1.7).

Fig. 1.6. Japanese bamboo vase which incorporates pentagonal and heptagonal rings. Courtesy Prof. Eiji Osawa.

1.3 Single-walled nanotubes

Nanotubes of the kind described by Iijima in 1991 invariably contain at least two graphitic layers, and generally have inner diameters of around 4 nm. In 1993, Iijima and Toshinari Ichihashi of NEC, and Donald Bethune and colleagues of the IBM Almaden Research Center in California independently reported the synthesis of *single-walled* nanotubes (1.13, 1.14). This proved to be an extremely important development, since the single-walled tubes appeared to have structures which approximate to those of the 'ideal' nanotubes shown in Fig. 1.2. Moreover, the single-walled tubes were completely novel. While multiwalled graphitic tubules, produced by catalysis, had been known for many years before the discovery of fullerene-related nanotubes (see next section), nothing like single-walled carbon nanotubes had been observed before. An early image of a single-walled nanotube (SWNT) sample is shown in Fig. 1.8(a). It can be seen that the appearance is quite different to that of samples of multiwalled nanotubes (MWNT). The individual tubes have very small diameters (typically ∼ 1 nm), and are curled and looped rather than straight. In this image, the tubes are contaminated with amorphous carbon and catalytic particles; subsequent work has enabled much purer samples to be produced. A higher magnification image of some individual tubes is shown in Fig. 1.8(b).

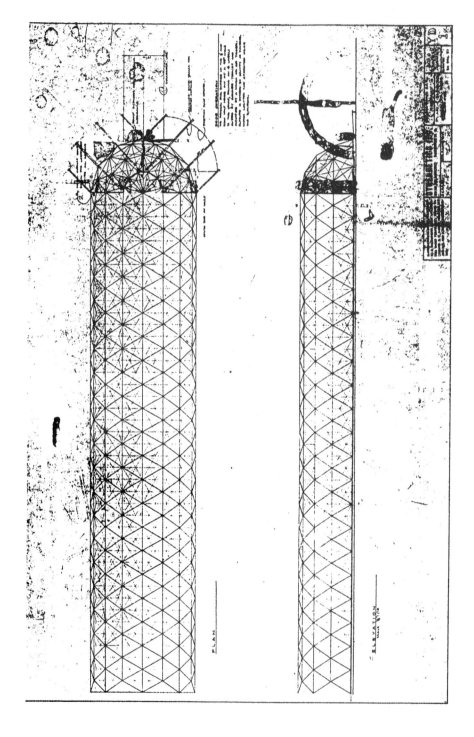

Fig. 1.7. The original 'buckytube'? Buckminster Fuller's design for the entrance pavilion of the Union Tank Car Company dome, Baton Rouge, Louisiana (1.12).

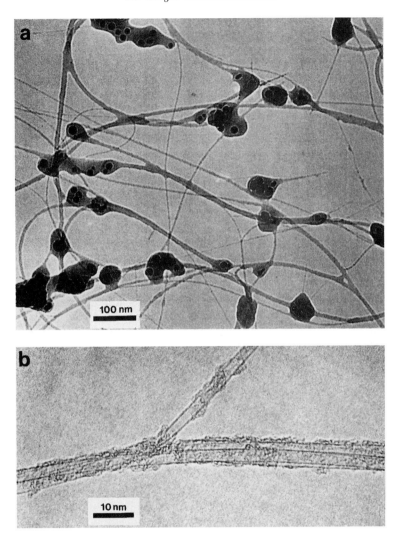

Fig. 1.8. Typical sample of single-walled nanotubes. (a) General area, showing tubes coated with amorphous carbon. Catalytic metal particles can also be seen (courtesy Donald Bethune). (b) Higher magnification image of individual tubes (1.15).

An alternative method of preparing single-walled nanotubes was described by Smalley's group in 1996 (1.16). Like the original method of preparing C_{60}, this involved the laser-vaporisation of graphite, and resulted in a high yield of single-walled tubes with unusually uniform diameters. These highly uniform tubes had a greater tendency to form aligned bundles than those prepared using arc-evaporation, and led Smalley to christen the bundles nanotube 'ropes'. Initial experiments indicated that the rope samples contained a very high proportion of nanotubes with a specific armchair structure. This created considerable excitement (1.17), since one of the problems with nanotube samples up to that point was the wide range of different structures present. Subsequent work has suggested that the rope samples may be less homogeneous than originally thought. Nevertheless, the synthesis of nanotube ropes gave an important boost to nanotube research, and some of the most impressive work has been carried out on these samples.

1.4 Pre-1991 evidence for carbon nanotubes

The discovery of carbon nanotubes was immediately recognised as an important event, and prompted the question: why were they not discovered many years ago? The same question has frequently been asked about buckminsterfullerene itself. All of the techniques required to prepare and characterise these materials have been available for decades, so why did we have to wait so long? In the case of carbon nanotubes, part of the answer seems to be that they *were* seen previously, but simply not recognised as a new and important form of carbon. From a historical point of view it is of interest to review the pre-1991 evidence for the existence of 'buckytubes', but this should not be seen as diminishing the importance of Iijima's work, which was the first to appreciate fully the nature and importance of these structures.

A letter in *Nature* in October 1992 claimed that nanotubes had been observed as early as 1953 (1.18). The writer described thread-like carbon structures obtained from the reaction of CO and Fe_3O_4 at 450 °C, which he suggested were 'similar, if not identical' to carbon nanotubes. This illustrates the care which must be taken in assessing claims of prior discovery in this area. Fine tubules of carbon produced by catalytic methods have been known for many years, but these have a much less perfect structure than the fullerene-related tubes, as discussed in Chapters 2 and 3. The tubes are not closed with fullerene-like caps, but usually have a metal particle attached to one end. Although there has been a substantial amount of research on carbon tubules produced by catalysis (e.g. 1.19), these imperfect structures do not lend themselves to the range of potential applications which seem possible for fullerene-related nanotubes.

A number of authors have also pointed out the similarities between carbon nanotubes and graphite whiskers, the highly perfect form of carbon fibres which were first prepared by Roger Bacon in 1960 (1.20). It is certainly true that the technique used to produce whiskers, which involves a DC carbon arc, seems very close to the arc-evaporation method of nanotube synthesis. However, there are major differences between graphite whiskers and carbon nanotubes, not least the fact that whiskers are far larger: typically about 5 μm in diameter and up to 3 cm in length. There are also structural differences: whiskers are scroll-like, while nanotubes are believed to have Russian-doll like structures (see Chapter 6 for further details on whiskers).

If catalytically produced tubes and graphite whiskers can be dismissed, what is the evidence that *genuine* carbon nanotubes were observed prior to 1991? To begin, there are Iijima's studies of carbon films carried out in the late 1970s and early 1980s, which were mentioned above. For this work he prepared specimens of arc-evaporated carbon using an apparatus of the type commonly employed to make carbon support films for electron microscopy. The method he used would have differed slightly (but significantly) from the Krät-schmer–Huffman technique in that the chamber would have been evacuated rather than filled with a small pressure of helium. The resulting films were largely amorphous, but contained small, partially graphitized regions which contained some unusual structures. These structures included discrete graphitic particles apparently made up of concentric closed shells, tightly curved around a central cavity. One of these structures, reproduced as Figure 5(a) in his 1980 *Journal of Microscopy* paper (1.8), is clearly a nanotube, and Iijima confirmed its tubular nature using tilting experiments. But he did not explore these structures in detail, and suggested that the curved layers were probably due to sp^3 bonding, rather than, as we now believe, the presence of pentagonal rings.

There are other examples of nanotube-like structures scattered throughout the pre-1991 carbon literature (see, for instance, Fig. 10 in Ref. (1.21)). In some cases these structures might be contaminants on the carbon films used to support the samples (see p. 34). Evidence for the existence of elongated fullerene-like structures also came from high resolution electron microscopy studies of C_{60} crystals. In one of the first such studies, published in July 1991 (1.22), Su Wang and Peter Buseck of Arizona State University reported that fullerite crystals contained closed carbon cages both smaller and larger than C_{60}. The larger fullerenes had the projected shapes of elongated ellipsoids, which were estimated to contain roughly 130 carbon atoms. It is possible that such structures arose from the coalescence of adjacent C_{60} or C_{70} molecules in the crystal.

Work by other authors also anticipated the discovery of carbon nanotubes. For example, Patrick Fowler of Exeter University described theoretical studies of small cylindrical fullerene molecules in early 1990 (1.23). Two groups of American theoreticians, one at the Naval Research Laboratory, Washington DC (1.24), and one at the Massachusetts Institute of Technology (1.25) submitted papers on the electronic properties of fullerene tubes just a few weeks before Iijima's paper appeared in *Nature*. Finally, the highly imaginative

British chemist David Jones, under his pen-name Daedalus, also ruminated about rolled-up tubes of graphite in the *New Scientist* in 1986 (1.26). He suggested that molecules of this type might behave as extremely viscous supercritical gases at room temperature, which could be useful in calming the atmosphere at violent public demonstrations. Although this particular prophecy did not prove accurate, Daedalus has an excellent record in the field of fullerene research. In the 1960s he had predicted the existence of spherical carbon molecules, anticipating the discovery of fullerenes by nearly 20 years.

1.5 Nanotube research

The method for producing nanotubes described by Iijima in 1991 gave relatively poor yields, making further research into their structure and properties difficult. A significant advance came in July 1992 when Thomas Ebbesen and Pulickel Ajayan, working at the same Japanese laboratory as Iijima, described a method for making gram quantities of nanotubes (1.27). Again, this was a serendipitous discovery: Ebbesen and Ajayan had been trying to make fullerene derivatives when they found that increasing the pressure of helium in the arc-evaporation chamber dramatically improved the yield of nanotubes formed in the cathodic soot. The availability of nanotubes in bulk gave an enormous boost to the pace of research worldwide.

One area which attracted early interest was the idea of using carbon nanotubes and nanoparticles as 'molecular containers'. A landmark in this field was the demonstration, by Ajayan and Iijima, that nanotubes could be filled with molten lead, and thus be used as moulds for 'nanowires' (1.28). Subsequently, more controlled methods of opening and filling nanotubes have been developed, enabling a wide range of materials, including biological ones, to be placed inside. The resulting opened or filled tubes might have fascinating properties, with possible applications in catalysis, or as biological sensors. Filled carbon nanoparticles may also have important applications, in areas as diverse as magnetic recording and nuclear medicine.

Perhaps the largest volume of research into nanotubes has been devoted to their electronic properties. The theoretical work which pre-dated Iijima's discovery has already been mentioned. A short time after the publication of Iijima's 1991 letter in *Nature*, two other papers appeared on the electronic structure of carbon nanotubes (1.29, 1.30). The MIT group, and Noriaki Hamada and colleagues from Iijima's laboratory in Tsukuba, carried out band structure calculations on narrow tubes using a tight-binding model, and demonstrated that electronic properties were a function of both tube structure and diameter. These remarkable predictions stimulated a great deal of interest, but attempting to determine the electronic properties of nanotubes experimentally presented great difficulties. Since 1996, however, experimental measurements have been carried out on individual nanotubes, which appear to confirm the theoretical predictions. These results have prompted speculation that nanotubes might become components of future nanoelectronic devices.

Determining the mechanical properties of carbon nanotubes also presented formidable difficulties, but once again experimentalists have proved equal to the challenge. Measurements carried out using transmission electron microscopy and atomic force microscopy have demonstrated that the mechanical characteristics of carbon nanotubes may be just as exceptional as their electronic properties. As a result, there is growing interest in using nanotubes in composite materials.

A variety of other possible applications of nanotubes are currently exciting interest. For example, a number of groups are exploring the idea of using nanotubes as tips for scanning probe microscopy. With their elongated shapes, pointed caps and high stiffness, nanotubes would appear to be ideally suited for this purpose, and initial experiments in this area have produced some extremely impressive results. Nanotubes have also been shown to have useful field emission properties, which might lead to their being used in flat-panel displays. Overall, the volume of nanotube research is growing at an astonishing rate, and commercial applications will surely not be far behind.

1.6 Organisation of the book

The chapter which follows considers the various methods for synthesising nanotubes, including catalytically produced and single-walled tubes, and summarises current thinking on growth mechanisms. Methods of purifying, aligning and processing nanotubes are also covered. In Chapter 3, theoretical approaches to the analysis of nanotube structure are outlined, and experimental observations described. The structures of carbon nanoparticles and nanocones are also discussed. Chapter 4, entitled 'The physics of nanotubes', is primarily concerned with the electronic properties of carbon nanotubes, but also considers magnetic, optical and vibrational properties of nanotubes, as well as experimental studies of nanotubes as field emitters. Methods of opening and filling carbon nanotubes and nanoparticles are described in Chapter 5, together with the possible uses of these filled nanocapsules. Theoretical and experimental work on the mechanical properties of nanotubes, and nanotube-containing composites are discussed in Chapter 6.

Chapter 7 covers inorganic analogues of fullerenes and nanotubes. Curved and tubular inorganic crystals have been known for many years, but there has been renewed interest in such structures since the discovery of 'inorganic fullerenes' based on dichalcogenides by Israeli workers in 1992. Fullerene-like structures containing boron and nitrogen are also discussed in this chapter. In Chapter 8 a discussion is given of spheroidal forms of carbon. The first part of the chapter is concerned with the recently discovered fullerene-like structures known as 'onions', and this is followed by a brief survey of more well established spheroidal carbon structures including soot and carbon black particles. The evidence that these structures contain fullerene-like elements is reviewed. Finally, Chapter 9 considers some possible future directions in which nanotube science might develop.

References

(1.1) H. W. Kroto, J. R. Heath, S. C. O'Brien, R. F. Curl and R. E. Smalley, 'C_{60}: Buckminsterfullerene', *Nature*, **318**, 162 (1985).

(1.2) J. Baggott, *Perfect symmetry: the accidental discovery of buckminsterfullerene*, Oxford University Press, 1994.

(1.3) H. Aldersey-Williams, *The most beautiful molecule*, Aurum Press, London, 1995.

(1.4) H. W. Kroto, 'Symmetry, space, stars and C_{60}' (Nobel lecture), *Rev. Mod. Phys.*, **69**, 703 (1997).

(1.5) R. E. Smalley, 'Discovering the fullerenes' (Nobel lecture), *Rev. Mod. Phys.*, **69**, 723 (1997).

(1.6) W. Krätschmer, L. D. Lamb, K. Fostiropoulos and D. R. Huffman, 'Solid C_{60}: a new form of carbon', *Nature*, **347**, 354 (1990).

(1.7) M. Ge and K. Sattler, 'Scanning tunnelling microscopy of single-shell nanotubes of carbon', *Appl. Phys. Lett.*, **65**, 2284 (1994).

(1.8) S. Iijima, 'High resolution electron microscopy of some carbonaceous materials', *J. Microscopy*, **119**, 99 (1980).

(1.9) S. Iijima, 'Direct observation of the tetrahedral bonding in graphitized carbon black by high-resolution electron microscopy', *J. Cryst. Growth*, **50**, 675 (1980).

(1.10) S. Iijima, 'Helical microtubules of graphitic carbon', *Nature*, **354**, 56 (1991).

(1.11) S. Iijima, T. Ichihashi and Y. Ando, 'Pentagons, heptagons and negative curvature in graphitic microtubule growth', *Nature*, **356**, 776 (1992).

(1.12) J. Ward, ed., *The artifacts of Buckminster Fuller*, *Vol.* 3, Garland Publishing, New York, 1985.

(1.13) S. Iijima and T. Ichihashi, 'Single-shell carbon nanotubes of 1-nm diameter', *Nature*, **363**, 603 (1993).

(1.14) D. S. Bethune, C. H. Kiang, M. S. de Vries, G. Gorman, R. Savoy, J. Vasquez and R. Beyers, 'Cobalt-catalysed growth of carbon nanotubes with single-atomic-layer walls', *Nature*, **363**, 605 (1993).

(1.15) C. H. Kiang, P. H. M. van Loosdrecht, R. Beyers, J. R. Salem, D. S. Bethune, W. A. Goddard III, H. C. Dorn, P. Burbank and S. Stevenson, 'Novel structures from arc-vaporized carbon and metals: single-layer nanotubes and metallofullerenes', *Surf. Rev. Lett.*, **3**, 765 (1996).

(1.16) A. Thess, R. Lee, P. Nikolaev, H. Dai, P. Petit, J. Robert, C. Xu, Y. H. Lee, S. G. Kim, A. G. Rinzler, D. T. Colbert, G. E. Scuseria, D. Tománek, J. E.

Fischer and R. E. Smalley, 'Crystalline ropes of metallic carbon nanotubes', *Science*, **273**, 483 (1996).

(1.17) P. Ball, 'The perfect nanotube', *Nature*, **382**, 207 (1996).

(1.18) J. A. E. Gibson, 'Early nanotubes?', *Nature*, **359**, 369 (1992).

(1.19) R. T. K. Baker and P. S. Harris, 'The formation of filamentous carbon', *Chem. Phys. Carbon*, **14**, 83 (1978).

(1.20) R. Bacon, 'Growth, structure and properties of graphite whiskers', *J. Appl. Phys.*, **31**, 283 (1960).

(1.21) G. R. Millward and D. A. Jefferson, 'Lattice resolution of carbons by electron microscopy', *Chem. Phys. Carbon*, **14**, 1 (1978).

(1.22) S. Wang and P. R. Buseck, 'Packing of C_{60} molecules and related fullerenes in crystals: a direct view', *Chem. Phys. Lett.*, **182**, 1 (1991).

(1.23) P. W. Fowler, 'Carbon cylinders: a new class of closed-shell clusters', *J. Chem. Soc., Faraday Trans.*, **86**, 2073 (1990).

(1.24) J. W. Mintmire, B. I. Dunlap and C. T. White, 'Are fullerene tubules metallic?', *Phys. Rev. Lett.*, **68**, 631 (1992).

(1.25) M. S. Dresselhaus, G. Dresselhaus and R. Saito, 'Carbon fibers based on C_{60} and their symmetry', *Phys. Rev. B*, **45**, 6234 (1992).

(1.26) D. E. H. Jones (Daedalus), *New Scientist*, **110** (1505), 88 and (1506), 80 (1986).

(1.27) T. W. Ebbesen and P. M. Ajayan, 'Large-scale synthesis of carbon nanotubes' *Nature*, **358**, 220 (1992).

(1.28) P. M. Ajayan and S. Iijima, 'Capillarity-induced filling of carbon nanotubes', *Nature*, **361**, 333 (1993).

(1.29) R. Saito, M. Fujita, G. Dresselhaus and M. S. Dresselhaus, 'Electronic structure of graphene tubules based on C_{60}', *Phys. Rev. B*, **46**, 1804 (1992).

(1.30) N. Hamada, S. Sawada and A. Oshiyama, 'New one-dimensional conductors: graphitic microtubules', *Phys. Rev. Lett.*, **68**, 1579 (1992).

2

Synthesis

Preparation methods, growth mechanisms and processing techniques

> . . . everything, both structure and change, is the outcome of chance
> orchestrations of chaos . . .
>
> P. W. Atkins, *The Creation*

The formation of beautifully symmetric C_{60} molecules in the chaotic high-temperature plasma of a carbon arc seems an almost magical process, and it is one which remains poorly understood. A number of models have been put forward, none of which has gained universal acceptance (2.1–2.5). The earliest theory of fullerene formation is the 'pentagon road model' put forward by the discoverers of C_{60} in 1987 (2.1). This assumes that when the cluster size reaches about 30 atoms, graphitic sheets begin to form. As further atoms are added, the formation of pentagonal rings will be favoured in order to minimise dangling bonds, and closure will occur 'accidentally', as a result of a fortuitous distribution of pentagons. However, it was realised that this simple model could not account for the high yield of C_{60} observed experimentally. A more sophisticated version of the 'pentagon road model' was therefore introduced (2.2), which stated that a growing carbon sheet will tend to incorporate the *maximum* possible number of pentagons, consistent with the proviso that the pentagons are isolated. The smallest closed structure which can be formed is then C_{60}. It was also emphasised that efficient production of C_{60} by the pentagon road mechanism requires sufficient annealing, so that all open clusters would have time to rearrange into the favoured 'pentagon rule' structure.

An alternative mechanism, known as the 'fullerene road model' was proposed by Jim Heath in 1991 (2.3). Here, it is assumed that the intermediates are closed cages containing 30–58 carbon atoms, rather than open fragments. Because these small fullerenes must contain adjacent pentagons, they are relatively reactive and therefore will tend to coalesce to form more stable fullerenes. Once C_{60}, C_{70} and other isolated pentagon fullerenes form they will

16

become end-points of cluster growth. However, this mechanism has problems explaining certain experimental observations, such as the formation of endohedral metallofullerenes. A number of other mechanisms of C_{60} formation have been put forward, none of which are without their shortcomings (2.4, 2.5). At present, the pentagon road mechanism probably remains the most popular explanation for fullerene formation in the arc.

The formation of carbon nanotubes on the cathode during arc-evaporation raises as many questions as those posed by fullerene synthesis. For example, why are nanotubes only seen in the carbon which forms on the cathode and not in the soot deposited on the walls of the vessel? What is the role, if any, of the electric field in nanotube formation? And why is the yield of nanotubes so sensitive to helium pressure? Although many theories have been put forward, most of these questions remain unanswered, and the uncertainty surrounding nanotube growth mechanisms has impeded progress towards development of more controlled synthesis techniques.

Of course, arc-evaporation is not the only technique for synthesising nanoscale graphitic tubes. As mentioned in the opening chapter, the production of relatively imperfect carbon tubes by catalytic methods has been known for decades, and in this case the mechanism appears to be rather better understood. A number of other methods for making both multiwalled and single-walled nanotubes have been introduced since 1991. Perhaps the most successful of these has been the laser vaporisation method for the synthesis of single-walled nanotubes which was introduced by Smalley's group in 1996 (2.6).

The first part of this chapter is concerned with the synthesis of multiwalled nanotubes. To begin, a description is given of the arc-evaporation method of nanotube synthesis. This is the 'classic' method of preparing multiwalled nanotubes, and produces the best quality samples. Alternative, non-catalytic, methods for producing nanotubes are also summarised. A brief discussion is then given of experiments on the high-temperature heat treatment of fullerene soot. Such treatments can result in the formation of carbon nanoparticles and nanotube-like structures and may provide insights into the mechanism of nanotube growth. Catalytic methods for preparing multiwalled nanotubes are then covered and, following some comments on the presence of nanotubes on TEM support grids, the synthesis of single-walled nanotubes is discussed.

After considering experimental methods of nanotube synthesis, an attempt is made to summarise current thinking on the mechanism of nanotube formation and growth, and to outline the main areas of uncertainty. Finally, a discussion is given of procedures which can be used to purify and align samples of nanotubes, and to control their lengths.

2.1 Production of multiwalled nanotubes: non-catalytic methods

2.1.1 The arc-evaporation technique

The original method used by Iijima to prepare nanotubes (2.7) differed slightly from the Krätschmer–Huffman technique for C_{60} production in that the graphite electrodes were held a short distance apart during arcing, rather than being kept in contact. Under these conditions, some of the carbon which evaporated from the anode re-condensed as a hard cylindrical deposit on the cathodic rod. It was the central part of this deposit that Iijima found to contain both nanotubes and nanoparticles. But the yield was rather poor in these initial experiments, as mentioned in Chapter 1, making further progress in the field initially rather slow. Subsequent modifications to the procedure, notably by Ebbesen and Ajayan (2.8) have enabled greatly improved yields to be obtained by arc-evaporation. The discussion which follows draws considerably on the excellent reviews of nanotube synthesis which have been given by Ebbesen (2.9, 2.10).

A variety of different arc-evaporation reactors have been used for nanotube synthesis, but a stainless steel vacuum chamber with a viewing port is probably the best type. A typical example is illustrated in Fig. 2.1. A glass-dome chamber of the kind used in the original Krätschmer–Huffman experiments is not ideal, since this does not easily allow for the separation of the rods to be adjusted during evaporation. The chamber must be connected both to a vacuum line with a diffusion pump, and to a helium supply. A continuous flow of helium at a given pressure is usually preferred over a static atmosphere of the gas. The electrodes are two graphite rods, usually of high purity, although there is no evidence that exceptionally pure graphite is necessary. Typically, the anode is a long rod approximately 6 mm in diameter and the cathode a much shorter rod 9 mm in diameter. Efficient water-cooling of the cathode has been shown to be essential in producing good quality nanotubes, and the anode is also frequently cooled. The position of the anode should be adjustable from outside the chamber, so that a constant gap can be maintained during arc-evaporation. A voltage-stabilised DC power supply is normally used, and discharge is typically carried out at a voltage of 20 V. The current depends on the diameter of the rods, their separation, the gas pressure and so on, but is usually in the range 50–100 A.

When the pressure is stabilised, the voltage should be turned on. At the start of the experiment the electrodes should not be touching, so no current will initially flow. The movable anode is now gradually moved closer to the cathode until arcing occurs. When a stable arc is achieved, the gap between the rods should be maintained at approximately 1 mm or less; the rod is

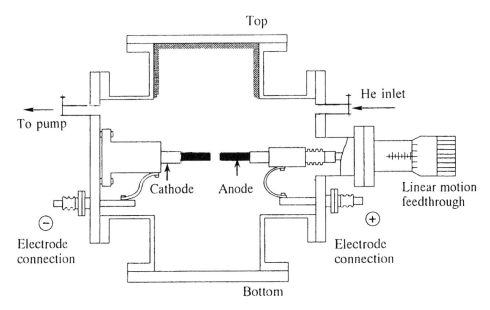

Fig. 2.1. Schematic illustration of arc-evaporation apparatus for the production of fullerenes and nanotubes (adapted from Ref. (2.11)). Although not shown here, it is usual for the electrodes to be water-cooled.

normally consumed at a rate of a few millimetres per minute. When the rod is consumed, the power should be turned off and the chamber left to cool before opening.

A number of factors have been shown to be important in producing a good yield of high quality nanotubes. Perhaps the most important is the pressure of the helium in the evaporation chamber, as demonstrated by Ebbesen and Ajayan in their 1992 paper (2.8). This is illustrated graphically in Fig. 2.2, taken from this paper, which shows nanotube samples prepared at 20, 100 and 500 Torr. A striking increase in the number of tubes is evident as the pressure is increased. At pressures above 500 Torr there is no obvious change in sample quality, but a fall in total yield. Thus, 500 Torr appears to be the optimum helium pressure for nanotube production. Note that these conditions are *not* optimum for C_{60} production, which requires a pressure of below 100 Torr.

Another important factor in the arc-evaporation method is the current, as demonstrated in several studies (2.12, 2.13). Too high a current will result in a hard, sintered material with few free nanotubes. Therefore, the current should be kept as low as possible, consistent with maintaining a stable plasma. Efficient cooling of the electrodes and the chamber has also been shown to be essential in producing good quality nanotube samples and avoiding excessive sintering, as discussed in the next section.

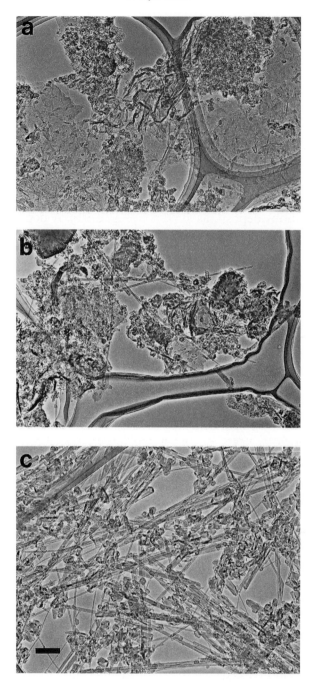

Fig. 2.2. Micrographs showing effect of helium on yield of nanotubes in arc-evaporation experiments, from the work of Ebbesen and Ajayan (2.8). Samples prepared at (a) 20 Torr, (b) 100 Torr and (c) 500 Torr.

2.1.2 The quality of nanotube samples produced by arc-evaporation

The shape and composition of the deposit which forms on the cathode following arc-evaporation depends strongly on the conditions used. A number of studies of the cathodic deposit have been made using optical microscopy and scanning electron microscopy (e.g. 2.12–2.14), and have produced widely differing results. However, it seems clear that the macroscopic structure of the deposit is particularly dependent on the efficiency of cooling. Thus, poor cooling results in a layered deposit as shown in Fig. 2.3(a). The nanotubes in such a deposit are found in small pockets, and tend to be randomly oriented. Efficient cooling of the electrodes, on the other hand, produces a more cylindrical and homogenous deposit, as shown in Fig. 2.3(b). A deposit of this kind consists of a hard outer shell, consisting of fused material, and a softer fibrous core which contains discrete nanotubes and nanoparticles. These can be extracted by cutting open the outer shell. Some indication of the quality of the nanotube samples can be gained by a simple physical examination of the carbon. A poor sample containing few nanotubes will generally have a powdery texture, while good quality material can be smeared to produce sheet-like flakes with a grey metallic lustre (note, however, that gloves should be worn when handling the carbon, as discussed below).

Examination of the fibrous core material by scanning electron microscopy has shown that it contains aligned microfibrils made up of nanotube bundles and individual nanotubes. However, the alignment of the nanotubes themselves is less obvious than for the much larger microfibrils. The ratio of nanotubes to nanoparticles in the best samples is of the order of 2:1. Samples of the nanotube-containing material can be prepared for electron microscopy by dispersing in a solvent such as isopropyl alcohol, ultasonicating and depositing onto carbon film support grids. It is important to note, however, that nanotubes and other graphitic materials can be present as *contaminants* on carbon film supports (see Section 2.4).

2.1.3 Safety considerations

Opinions are divided on the possible health hazards of carbon nanotubes (2.15, 2.16). Some workers have pointed out the physical similarities between nanotubes and asbestos fibres, which are both extended structures typically around 10 nm in diameter and a few micrometres long (see Chapter 7 for further details on asbestos fibres). It has been known since the early 1960s that asbestos can be a cause of pneumoconiosis, a serious lung disease, and mesothelioma, a cancer of the lining of the chest which is often fatal. However, the

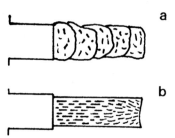

Fig. 2.3. Sketches of deposits formed on cathode by arc-evaporation (2.9). (a) Deposit with layered structure, (b) homogeneous deposit with aligned bundles of nanotube-containing material.

mechanism by which the silicate fibres cause the damage, at least in the case of mesothelioma, is believed to involve the catalytic formation of reactive oxygen compounds. It seems unlikely that nanotubes would have the same effect. Nevertheless, in the lack of any definite information on the toxicity of fullerene-related carbons, it is wise to err on the side of caution when preparing and handling these materials. Particular care should be taken with the arc-evaporation method since the soot produced in this way is extremely light and can easily become airborne. Precautions should therefore be taken to avoid inhalation. For this reason, it is recommended that the entire arc-evaporation apparatus is enclosed in a fume hood. A mask should also be worn when opening the chamber, and it is advisable to wear gloves when handling the fullerene-related materials.

There are other safety considerations to take into account when carrying out the arc-evaporation method. It is clearly important to check the machine for short circuits before carrying out arc-evaporation, and the vacuum should be tested for leaks before introducing the inert gas. Since most chambers will have a viewing port, care must also be taken to protect the operator's eyes from the intense light of the arc using a high density optical glass filter.

2.1.4 Condensation of carbon vapour in the absence of an electric field

A quite different method of nanotube production was described by a group at the Russian Academy of Sciences led by Leonid Chernozatonskii, in 1992 (2.17). These workers used an electron beam to evaporate graphite in a high vacuum (10^{-6} Torr), and collected the material which condensed on a substrate such as quartz. The condensed material appeared to consist of a mass of aligned fibres, which the authors suggested were carbon nanotubes. At first this work was greeted with some scepticism, mainly because the material was characterised by relatively low resolution scanning electron microscopy, which did not provide clear evidence that the structure was made up of nanotubes rather than ordinary graphite. However, subsequent examination of the samples using high resolution electron microscopy (2.18) apparently revealed the presence of rather imperfect multiwalled nanotubes in the condensed film.

Nanotube production by the vapour-condensation of carbon was also described by Maohui Ge and Klaus Sattler of the University of Hawaii (2.19, 2.20). In this case, carbon vapour was produced by resistively heating a carbon foil, and the vapour was condensed onto freshly cleaved highly oriented pyrolytic graphite (HOPG), under a vacuum of 10^{-8} Torr. It was claimed that both multiwalled and single-walled nanotubes could be produced in this way.

In 1995, Smalley's group at Rice University reported the synthesis of carbon nanotubes by oven laser-vaporisation (2.21). In this technique, originally developed to study the effect of carrier gas temperature on C_{60} production, a laser is used to vaporise a graphite target held in a controlled-environment oven, as shown in Fig. 2.4. The carrier gas used was helium or argon, and the oven temperature was approximately 1200 °C. The condensed material was collected on a cooled target as shown in the diagram, and was found to contain a significant proportion of nanotubes and nanoparticles, which appeared to be highly graphitized and structurally perfect. Subsequent development of the oven laser-vaporisation method by Smalley's group proved to be of great value in the production of single-walled nanotubes, as discussed in Section 2.5.3.

2.1.5 Pyrolytic methods

Studies by Morinobu Endo and colleagues from Shinshu University, together with workers from Sussex, have shown that multiwalled nanotubes can be produced by the pyrolysis of benzene in the presence of hydrogen (2.22–2.24). Their method involved introducing benzene vapour and hydrogen into a ceramic reaction tube in which a central graphite rod was positioned to act as a substrate. The temperature was raised to 1000 °C and held at this level for 1 hour before cooling to room temperature and flushing with argon. The deposited material was then scraped from the substrate and subjected to a 'graphitizing' heat treatment at 2500–3000 °C for about 10 minutes. Examination of the resulting material in the TEM showed the presence of multiwalled nanotubes apparently similar in structure and quality to those produced by arc-evaporation. This method would therefore seem to have some potential for the bulk production of multiwalled tubes, although at present the yields appear to be low.

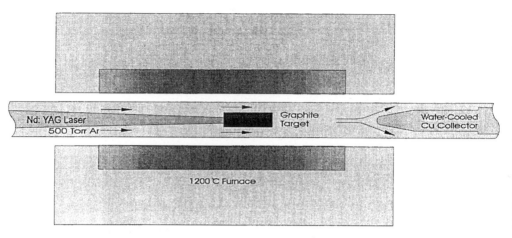

Fig. 2.4. Schematic illustration of the Smalley group's oven laser-vaporisation apparatus for the synthesis of multiwalled nanotubes and nanoparticles (2.21).

2.1.6 *Electrochemical synthesis of nanotubes*

An electrochemical method for the synthesis of multiwalled nanotubes has been described by the Sussex group (with the present author) (2.25, 2.26). This involved the electrolysis of molten lithium chloride using a graphite cell in which the anode was a graphite crucible and the cathode a graphite rod immersed in the melt. A current of about 30 A was passed through the cell for 1 minute, after which the electrolyte was allowed to cool, and then added to water to dissolve the lithium chloride and react with the lithium metal. The mixture was left for 4 hours, and toluene was then added to the aqueous suspension and the whole agitated for several minutes.

This treatment resulted in most of the solid material passing into the toluene layer, which was then separated from the aqueous layer by decanting. A few drops of the toluene suspension were then dropped onto a carbon film for examination by transmission electron microscopy. The material was found to contain large numbers of rather imperfect multiwalled nanotubes which were similar in appearance to catalytically formed tubes, as can be seen in Fig. 2.5. Nanoparticles were also found, and both tubes and nanoparticles often contained encapsulated material, presumably lithium chloride or oxide. Thus, the technique could prove to be a useful method of preparing filled carbon nanostructures.

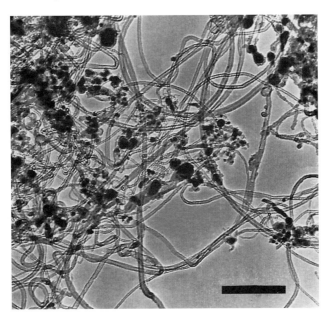

Fig. 2.5. Typical appearance of carbon nanotubes produced by electrolysis (courtesy Mauricio Terrones).

2.2 Experiments on the heat treatment of fullerene soot

As noted in the introduction to this chapter, experiments on the heat treatment of fullerene soot and other microporous carbons may provide insights into the nucleation and growth of multiwalled nanotubes, so it is appropriate to give a brief discussion of this topic here.

Fullerene soot is the light, fluffy carbon which forms on the walls of the evaporation vessel during fullerene synthesis. It is this soot which contains the C_{60}, C_{70} and higher fullerenes, which can be extracted using organic solvents. A typical micrograph of a sample of fullerene soot from which the fullerenes have been removed is shown in Fig. 2.6. It can be seen that the structure is highly disordered, consisting mainly of randomly curved graphene layers. In general, the layers appear to be very tightly curled, enclosing pores less than 2 nm in size, although sometimes larger, randomly shaped pores are present as can be seen in Fig. 2.6. The material is believed to consist of fragments of fullerene-like carbon in which both pentagons and heptagons are distributed randomly throughout a hexagonal network, producing continuous curvature (2.27–2.29), as in the structure shown in Fig. 2.7. This structure resembles the 'random schwarzite' structure proposed by Townsend *et al.* (2.30), although with many fewer seven-membered rings. The size of the individual fragments is

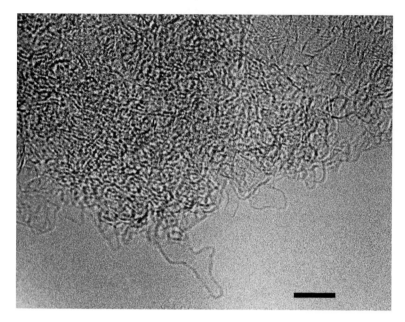

Fig. 2.6. High resolution electron micrograph of fullerene soot. Scale bar 5 nm.

Fig. 2.7. Schematic illustration of a model for the structure of fullerene soot, consisting of randomly curved, fullerene-like fragments.

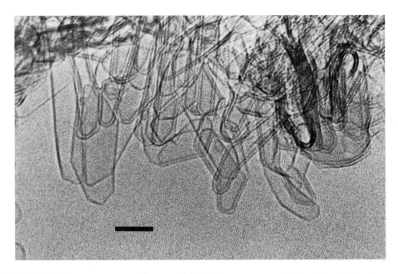

Fig. 2.8. Tube-like structures observed in fullerene soot following high-temperature heat treatment in a positive-hearth electron gun (2.28). Scale bar 5 nm.

not known, and cannot easily be determined using HREM, but is probably relatively small.

High temperature heat treatments of fullerene soot were first carried out by the Oxford group (2.27, 2.28) and by Walt de Heer and Daniel Ugarte from the Ecole Polytechnique Fédérale de Lausanne in Switzerland (2.31, 2.32). The results were somewhat different. The Oxford group found that heat treatment produced a structure apparently made up of large pores which were often extended in shape, resembling large-diameter single-layer nanotubes, as can be seen in the micrograph shown in Fig. 2.8. Like nanotubes, the extended pores were almost invariably closed, and exhibited a variety of capping morphologies. In some cases features were observed which are thought to be indicative of the presence of seven-membered carbon rings. The extended pores were usually bounded by single carbon layers, although multilayer structures were also present.

De Heer and Ugarte found that high-temperature heat treatments tended to transform the fullerene soot into small, graphitic nanoparticles rather than single-walled nanotubes. A series of micrographs taken from their work which illustrates this transformation is shown in Fig. 2.9 (2.31). Occasionally, rather short multiwalled nanotubes were observed in their heat-treated soot, but extended tubes like those formed by arc-evaporation were not seen.

These observations led the present author and his colleagues to propose a model for nanotube formation in which the tubes were envisaged to form as a result of high-temperature annealing of fullerene soot deposited onto the

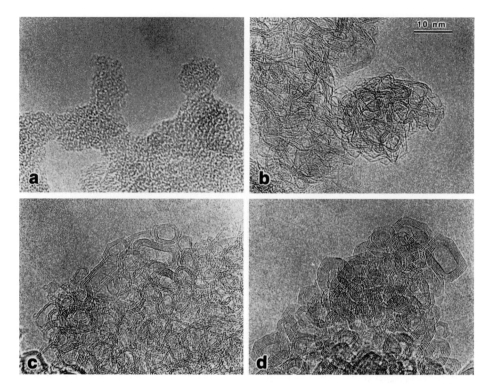

Fig. 2.9. Micrographs by De Heer and Ugarte (2.31) showing effect of high-temperature heat treatment on fullerene soot. (a) Crude soot, (b) heated at 1700 °C for 18 h, (c)
1700 °C (1 h) and 2100 °C (1 h), (d) 1700 °C (1 h) and 2250 °C (1 h).

cathode, rather than directly from the vapour phase. The model will be
discussed further in Section 2.6.4.

It is worth noting that conventional microporous carbons, prepared by the
pyrolysis of organic materials, can also be converted into a structure containing carbon nanoparticles by high-temperature heat treatment (2.33). This may
indicate that the microporous carbons have structures very similar to that of
fullerene soot, as discussed in a recent review (2.34).

2.3 Catalytically produced multiwalled nanotubes

2.3.1 Background

It has been known for over a century that filamentous carbon can be formed by
the catalytic decomposition of a carbon-containing gas on a hot surface. The
phenomenon was first observed by P. and L. Schultzenberger in 1890, during
experiments involving the passage of cyanogen over red-hot porcelain (2.35).

Work in the 1950s established that filaments could be produced by the interaction of a wide range of hydrocarbons and other gases with metals, the most effective of which were iron, cobalt and nickel. In all cases filament growth was found to be enhanced by the presence of hydrogen. Serious research into the catalytic formation of carbon filaments began in the 1970s when it was appreciated that filament growth could constitute a serious problem in certain chemical processes, as well as in the operation of gas-cooled nuclear reactors. This research, which has been reviewed in detail by Baker and Harris (2.36), focused on carbon deposition from two sources: the disproportionation of carbon monoxide or the decomposition of hydrocarbons. The CO disproportionation reaction can be represented by the Boudouard equilibrium:

$$2CO_{(g)} \Leftrightarrow C_{(s)} + CO_{2(g)}.$$

The maximum rates of carbon deposition from this reaction occur at temperatures of around 550 °C in the presence of metal particles of the iron subgroup. Transmission electron microscope studies of filaments produced in this way have shown that they are usually hollow, and can have three basic morphologies: helical, twisted or straight. The filaments have diameters ranging from 10 nm to 0.5 μm and can be up to 100 μm in length. Some TEM images of typical catalytically formed nanotubes are shown in Fig. 2.10.

The work of Baker and his colleagues was primarily motivated by the need to *avoid* filamentous carbon growth in situations such as the cooling circuits of nuclear reactors. However, it was recognised that the filamentous carbon produced in this way could constitute a new form of commercially useful carbon fibre. Over the past ten years or so there have been increasing efforts to develop a commercial process for the production of carbon fibres using catalytic pyrolysis, most notably by Morinobu Endo and his colleagues in Japan (2.38, 2.39). Much of their work has involved the controlled decomposition of benzene on a catalytic substrate. Thus, in a typical experiment, high-purity hydrogen would be passed through benzene, and the resulting mixture would then flow across a catalytically treated substrate held in a furnace at an initial temperature of approximately 1000 °C. Under these conditions, fibres would nucleate and begin to grow on the tiny catalytic particles covering the substrate. Subsequently, the temperature would be increased in order to promote further growth and thickening of the fibres by direct decomposition of the benzene. Under these conditions, fibres with diameters of approximately 10 μm and lengths up to about 25 cm can be produced. An alternative method has also been used by Endo and co-workers to make 'vapour grown carbon fibres' (VGCF), which avoids the use of a substrate. In this 'volume seeding'

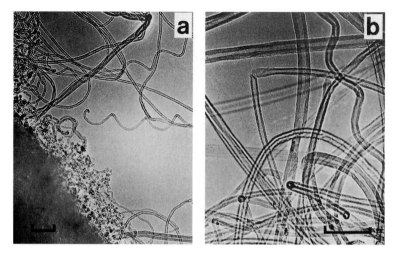

Fig. 2.10. (a), (b) Micrographs of carbon filaments formed by catalytic decomposition of acetylene on Co/SiO_2 (2.37).

technique, ultrafine catalyst particles are introduced into the feedstock so that fibres can grow in the three-dimensional space of the reactor, rather than just on a two-dimensional surface. This tends to produce rather thinner fibres than the substrate method, with diameters of around 1 μm.

Some TEM micrographs of vapour-phase-grown carbon fibres are shown in Fig. 2.11. These fibres were produced by the decomposition of benzene on iron particles. Like the filaments produced by Baker and his colleagues, the benzene-produced fibres are hollow, with a small catalytic particle remaining at the tip, but they are generally quite straight rather than curled or helical. The degree of graphitization in the catalytically grown fibres is rather low, but the fibres can be graphitized by heating at *c.* 3000 °C. The fibres produced in this way can have outstanding mechanical properties.

2.3.2 *Growth mechanisms of catalytically produced nanotubes*

Detailed insights into the mechanism of filament growth have come from the skilful application of controlled atmosphere electron microscopy (CAEM) by Baker and co-workers (e.g. 2.36), from about 1972 onwards. This work demonstrated directly for the first time that filament growth involved the deposition of carbon behind an advancing metal particle, with the forward face remaining apparently clean. The CAEM technique also enabled the kinetics of the process to be determined directly, showing that the activation energy for filament growth was about the same as the activation energy for bulk carbon

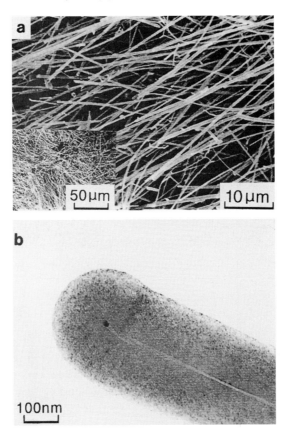

Fig. 2.11. Micrographs of 'vapour-phase-grown carbon fibres', from the work of Morinobu Endo and colleagues (2.39). (a) SEM micrographs of fibres formed by floating catalyst method, (b) TEM image showing fibre at early stage of growth. Iron catalyst particle can be seen near tip.

diffusion in nickel. This result, together with the direct observations made by CAEM, led Baker *et al.* to propose the mechanism for filament growth illustrated in Fig. 2.12. According to this mechanism, the first stage involves the decomposition of the hydrocarbon on the 'front' surface of the metal particle, producing hydrogen and carbon, which then dissolves in the metal. The dissolved carbon then diffuses through the particle, to be deposited on the trailing face, forming the filament. Direct deposition of carbon onto the filament can also occur, as shown in Fig. 2.12(c). This mechanism was originally applied to the growth of filaments on nickel in an atmosphere of acetylene, but it appears to apply more generally to most other systems.

The discovery of fullerene-related nanotubes in 1991 stimulated a renewed burst of fundamental work on the growth and structure of catalytically grown

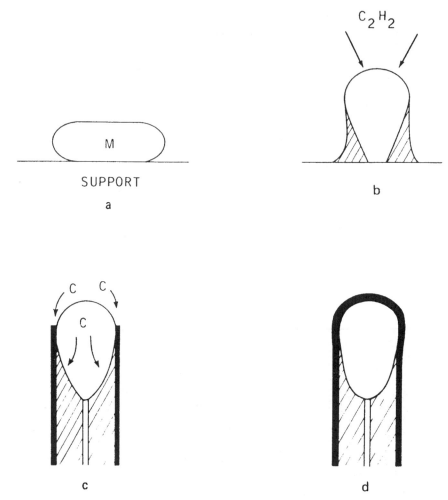

Fig. 2.12. Model proposed by Baker and Harris for the growth of catalytic carbon filaments (2.36).

carbon filaments. By drawing partly on the knowledge gained from nanotube research, these studies have led to some significant new insights. Particularly notable has been the work of Severin Amelinckx and colleagues at the University of Antwerp on curled and helical nanotubes (2.40, 2.41). Some of their images of helically coiled tubes are shown in Figs. 2.13 and 2.14. Amelinckx *et al.* discuss the growth of these tubes in terms of a locus of active sites around the periphery of the catalytic particle, and growth velocity vectors. In the simplest case the locus of active sites is circular and the extrusion velocity is constant, producing a straight tube propagating at a constant rate. If the catalytic activity varies around the circle, such that the velocity vectors termin-

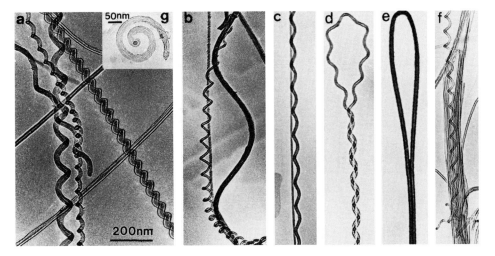

Fig. 2.13. Examples of helical nanotubes grown by the cobalt-catalysed decomposition of acetylene (2.41).

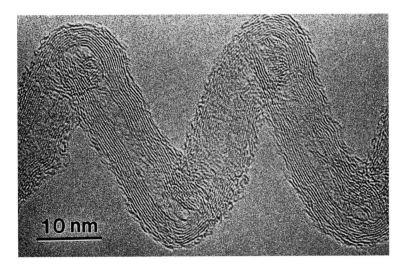

Fig. 2.14. High resolution micrograph of helical nanotube (2.41).

ate in a plane which is not parallel with the locus of active sites, a curved tube results. In practice, of course, the locus of active sites may not be circular, and this introduces another level of complexity. Amelinckx *et al.* consider the case of an elliptical locus of active sites. In this case, they show that a catalytic activity which varies around the ellipse will produce helical growth. A detailed discussion of their model is given in Ref. (2.40).

2.3.3 *Synthesis of aligned nanotubes by catalysis*

There are two approaches which can be used to produce aligned nanotubes. One is to carry out some kind of alignment procedure on previously prepared nanotubes, as discussed in Section 2.9. The other is to grow aligned tubes on substrates using catalytic methods. There is particular interest in the latter approach in connection with preparing nanotube-based field emission devices (see p. 149), and various methods have been described.

In 1996 a Chinese group described a technique in which iron nanoparticles were embedded in mesoporous silica and then used to catalyse the decomposition of acetylene at 700 °C (2.42). This resulted in the formation of straight multiwalled nanotubes growing in a direction perpendicular to the surface of the silica. It was reported that nanotube arrays of several square millimetres in area could be grown in this way, and that the arrays could be readily detached from the substrate. A slightly different method for the synthesis of aligned carbon nanotubes was described a short time later by Mauricio Terrones and colleagues (2.43). These workers used laser etching to prepare a patterned cobalt catalyst on a silica substrate, which was then heat treated to break up the Co films into discrete particles. The patterned catalyst was used to pyrolyse an organic compound at 950 °C, resulting in the formation of aligned nanotubes growing approximately parallel to the substrate. High resolution electron microscopy showed that the tubes were well graphitized.

Perhaps the most impressive demonstration of aligned nanotube growth on a substrate was given in late 1998 by a group from the State University of New York (2.44). This work was notable in being carried out at a relatively low temperature, which enabled the tubes to be grown on glass. The method involved firstly depositing a thin nickel layer onto the glass, and then using this as a catalyst to grow nanotubes by plasma-enhanced hot filament CVD, with acetylene as the carbon source. Aligned arrays of tubes were formed over several square centimetres. An SEM micrograph showing the excellent alignment achieved is shown in Fig. 2.15.

2.4 Nanotubes on TEM support grids: a word of warning

When carrying out TEM studies of carbon nanotubes it is essential to be aware that nanotubes and nanoparticles can sometimes be present as contaminants on evaporated carbon support films. Needless to say, this is particularly important in studies aimed at exploring new methods for nanotube synthesis, and there are examples in the literature where these contaminants have almost certainly been mistaken for sample material. As a result, several

Fig. 2.15. Scanning electron micrograph of aligned nanotubes grown on glass using catalytic methods (2.44).

papers claiming new methods for nanotube synthesis should be treated with caution.

In a recent study (2.45), the present author examined ten unused evaporated carbon films supported on copper grids, obtained from two different suppliers. Three types of carbon film were examined: 'lacey', 'holey' and continuous. Carbon nanotubes, nanoparticles or other graphitic carbon structures were found to be present as contaminants on all of the grids examined. The amount, and nature, of these contaminants varied considerably from grid to grid: in some cases large clusters of nanotubes and nanoparticles could be found quite easily, while in others only isolated nanoparticles and other structures were seen. Figure 2.16 shows a typical cluster of nanotube-containing material found on a commercial 'lacey' carbon film. Close examination of such regions showed them to be made up almost exclusively of graphitic nanotubes and nanoparticles. High resolution images showed that the tubes were invariably multilayered, and similar in appearance to those prepared by conventional arc-evaporation. However, the structure of the individual layers generally appeared somewhat less perfect than in tubes prepared by conventional arc-evaporation, and in many cases the tubes were found to have a thin surface coating of amorphous material. In addition

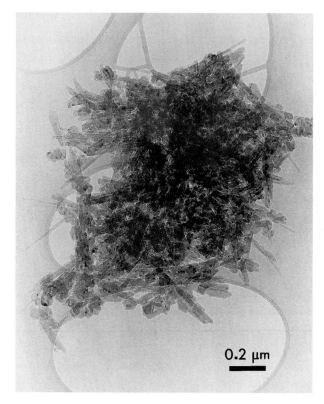

Fig. 2.16. Cluster of nanotubes found on commercial 'holey' carbon TEM support film (2.45).

to nanotubes and nanoparticles, other less-well-defined graphitic carbon structures are also frequently found on the support films (2.45, 2.46).

It is not too surprising that nanotubes and nanoparticles are present on carbon TEM support films since these are prepared by the arc-evaporation of graphite. The only significant difference between the preparation of carbon films and the synthesis of fullerenes and nanotubes is that in the latter case an atmosphere of helium is employed in the arc-evaporation vessel rather than a vacuum. It is also worth recalling that some of the earliest observations of nanotube-like structures were made by Iijima in studies of evaporated carbon films very similar to those used as support films (2.47, 2.48). The other graphitic structures observed on the carbon films are unlikely to be a result of arc-evaporation, but probably originate from an earlier stage of the production process, as discussed by Rietmeijer (2.46).

In view of these observations, it is advisable to examine TEM support films before using them for studies of nanotubes, in order to check their purity. Alternatively, non-carbon support films, such as those made from silicon monoxide, could be used, although these are not usually available in 'holey' form.

2.5 Single-walled nanotubes

2.5.1 Discovery

In early 1993, several groups reported that foreign materials could be encapsulated inside carbon nanoparticles or nanotubes by carrying out arc-evaporation using modified electrodes. Rodney Ruoff's group in California (2.49) and Yahachi Saito's group in Japan (2.50) prepared encapsulated crystals of LaC_2 by employing electrodes impregnated with lanthanum, while Supapan Seraphin and colleagues reported that YC_2 could be introduced into nanotubes by using electrodes containing yttrium (2.51). This work opened the way to a whole new field based on the use of nanoparticles and nanotubes as 'molecular containers', as described in Chapter 5, but it also led indirectly to a quite different discovery, with equally important implications.

Donald Bethune and his colleagues of the IBM Almaden Research Center in San Jose, California were particularly interested in the papers of Ruoff and the others. This group were working on magnetic materials for applications in information storage, and believed that ferromagnetic transition metal crystallites encapsulated in carbon shells might be of great value in this area. In such materials, the enclosed metal particles would retain their magnetic moments while being chemically and magnetically isolated from their neighbours. For some years, the IBM group had been working on 'endohedral fullerenes': fullerenes containing a small number of metal atoms, but larger clusters or crystals inside fullerene-like shells might be of more practical interest. Bethune therefore set out to try some arc-evaporation experiments using electrodes impregnated with the ferromagnetic transition metals iron, cobalt and nickel. But the result of this experiment was not at all what he expected. To begin with, the soot produced by arc-evaporation was quite unlike the normal material produced by arc-evaporation of pure graphite. Sheets of soot hung like cobwebs from the chamber walls, while the material deposited on the walls themselves had a rubbery texture, and could be peeled away in strips. When Bethune and his colleague Robert Beyers examined this strange new material using high resolution electron microscopy they were astonished to find that it contained multitudes of nanotubes with *single-atomic-layer walls*. These ultrafine tubes were entangled with amorphous soot and particles of metal or metal carbide, holding the material together in a way that would account for its strange texture. This work was written up for *Nature* and appeared in June 1993 (2.52). Micrographs taken from their paper are shown in Fig. 2.17.

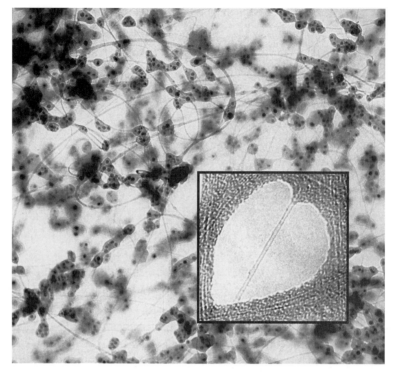

Fig. 2.17. Images from the work of Bethune *et al.* (2.52) showing single-walled carbon nanotubes produced by co-vaporisation of graphite and cobalt. The tubes have diameters of approximately 1.2 nm.

Independently of the American group, Sumio Iijima and Toshinari Ichihashi of the NEC laboratories in Japan were also experimenting with arc-evaporation using modified electrodes. In addition, they were interested in the effect of varying the atmosphere inside the arc-evaporation chamber. Like Bethune and colleagues, they discovered that certain conditions produced a quite different type of soot from that normally formed by arc-evaporation. For this work, the Japanese workers impregnated their electrodes with iron, and the atmosphere in the arc-evaporation chamber was a mixture of methane and argon rather than pure helium. When examined by high resolution electron microscopy, the arc-evaporated material was found to contain extremely fine single-walled nanotubes running like threads between clusters containing amorphous carbon and metal particles (2.53).

As noted in Chapter 1, single-walled nanotubes differ from those produced by conventional arc-evaporation in having a very narrow range of diameters. In the case of 'conventional' tubes, the inner diameter can range from *c.* 1.5 nm to *c.* 15.0 nm, and the outer diameter from *c.* 2.5 nm to *c.* 30 nm. The single-walled tubes, on the other hand, all have extremely narrow diameters. In the material produced by Bethune and colleagues, the tubes had diameters of 1.2 (± 0.1) nm, while Iijima and Ichihashi found that the tube diameters ranged from about 0.7 nm to 1.6 nm with the average being approximately 1.05 nm. Like tubes produced by conventional arc-evaporation, all the single-walled tubes appeared to be capped, and there was no evidence that catalytic metal particles were present at the ends of the tubes. Nevertheless, the growth of single-walled tubes is believed to be essentially catalytic, as discussed further in Section 2.6.5.

2.5.2 Subsequent work on single-walled tubes

Following their initial, ground-breaking work, Donald Bethune and his colleagues at IBM in San Jose, in collaboration with workers from the California Institute of Technology and the Virginia Polytechnic Institute and State University, have carried out a whole series of studies on the production of single-walled nanotubes using a variety of 'catalysts'. In one of the first of these (2.54), they showed that the addition of sulphur to the cobalt in the anode (either as elemental S, or as CoS), resulted in a much wider range of nanotube diameters than obtained from cobalt alone. Thus, single-walled nanotubes with diameters ranging from 1 to 6 nm were produced when sulphur was present in the cathode, compared to *c.* 1–2 nm for pure cobalt. It was subsequently shown that bismuth and lead could similarly promote the formation of large diameter tubes (2.55).

In 1997 a French group showed that high yields of single-walled nanotubes could be achieved with arc-evaporation (2.56). Their method was similar to the original technique of Bethune and colleagues, but with a slightly different reactor geometry. Also, the catalyst used was a nickel/yttrium mixture rather than the cobalt generally favoured by the Bethune group. The highest concentration of nanotubes was found to form in a 'collar' around the cathodic deposit, which made up approximately 20% of the total mass of evaporated material. Overall, the yield of tubes was estimated to be 70–90%. Examination of the 'collar' material by high resolution electron microscopy showed many bundles of tubes, with diameters around 1.4 nm. The yield and nature of the tubes produced were similar to the 'rope' samples synthesised by the Smalley group using laser vaporisation (see next section).

The production of single-walled nanotubes in a rather different form was described towards the end of 1993 by Shekhar Subramoney, of Du Pont in Wilmington, Delaware, in collaboration with workers from SRI International. These workers carried out arc-evaporation using gadolinium-impregnated electrodes (2.57), and collected the soot which condensed on the reactor walls. In addition to large amounts of amorphous carbon, the soot was found to contain 'sea-urchin' structures which proved to consist of single-walled nanotubes growing out of relatively large particles of gadolinium carbide (typically a few tens of nanometres in size). The tubes were generally much shorter than those produced by the iron group metals, but had a similar range of diameters. Subsequent work has shown that radial single-walled nanotubes can be formed on a variety of other metals including lanthanum and yttrium (2.58, 2.59). Figure 2.18, taken from the work of Saito and colleagues, shows a typical image of single-walled tubes growing radially from a lanthanum-containing particle. Unlike the iron group metals, the rare earth elements are not known as catalysts for the production of multiwalled nanotubes, so the formation of tubes on these elements is rather surprising, and the fact that the tubes grow on relatively large particles suggests that the mechanism is different. It has been suggested that the growth of tubes on the surfaces of the particles may involve segregation of supersaturated carbon atoms from the interior of carbide particles (2.59). It is worth noting that the radial growth of multiwalled tubes from catalytic particles was observed many years ago by Baker and others (e.g. 2.60).

The methods for producing single-walled nanotubes discussed so far have involved arc-evaporation using modified electrodes. Work by Richard Smalley and his colleagues from Rice University has shown that single-walled tubes can also be synthesised using a purely catalytic method (2.61). The catalyst employed contained particles of molybdenum a few nanometres in diameter, supported on alumina. This was placed inside a tube furnace through which carbon monoxide was passed at a temperature of 1200 °C. This temperature is rather higher than usually used for producing nanotubes by catalysis, and may explain why single-walled rather than multiwalled nanotubes were formed.

The catalytically produced single-walled tubes had a number of interesting features, which distinguished them from tubes synthesised by arc-evaporation. To begin with, the catalytic tubes generally had small metal particles attached to one end, as observed in multiwalled tubes produced by catalysis. There was also a relatively wide range of particle diameters (approximately 1–5 nm), and it appeared that the diameter of each tube was determined by that of the associated catalytic particle. Finally, the catalytically formed single-walled tubes were generally isolated rather than grouped into bundles as is frequently the case with the tubes synthesised by arc-evaporation.

These observations led Smalley and colleagues to suggest a growth mechanism for the catalytically formed tubes, which involves the initial formation of a single-layer cap (which they call a *yarmulke*, after the Yiddish name for a

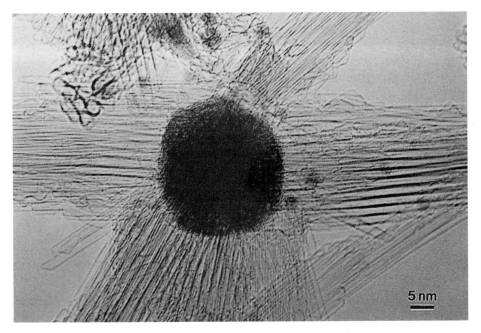

Fig. 2.18. Single-walled nanotubes growing on lanthanum particle (2.58).

skullcap), followed by the growth of this cap away from the catalytic particles, leaving a tube in its wake. This mechanism is quite different from the one they have proposed for the growth of single-walled tubes by laser vaporisation (see Section 2.6.5).

2.5.3 Nanotube 'ropes'

Since the discovery of C_{60} at Rice in 1985, Smalley's group has concentrated on the use of lasers to synthesise fullerene-related materials. The synthesis of multilayered nanotubes in this way has been described above (p. 23). In 1995 they reported a development of the laser synthesis technique which enabled them to prepare single-walled nanotubes in high yield (2.62). Subsequent refinements to this method led to the production of single-walled tubes with unusually uniform diameters (2.6). The apparatus used was similar to that shown in Fig. 2.4, but a composite metal–graphite target was used instead of pure graphite. The best yield of uniform single-walled nanotubes was obtained with a catalytic mixture comprising equal amounts of Co and Ni, and a double laser pulse was used to provide a more even vaporisation of the target.

Some micrographs of the material produced by this technique are shown in Fig. 2.19. In general appearance it is quite similar to the material produced by

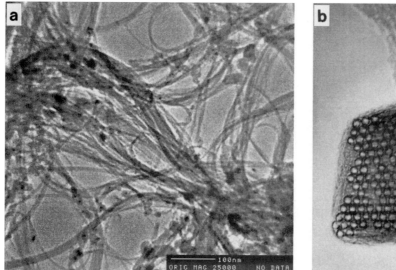
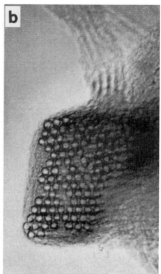

Fig. 2.19. TEM images of single-walled nanotube 'rope' samples (2.6). (a) Low magnification image showing a large number of ropes, (b) high resolution micrograph of individual rope imaged along axis.

arc-evaporation. However, the individual tubes apparently have a greater tendency to form 'ropes' or aligned bundles, which is consistent with the individual tubes having a uniform diameter. Ropes could sometimes be found which ran for a short distance in the beam direction, enabling them to be imaged 'end-on', as in Fig. 2.19(b). In addition to electron microscopy, Smalley and colleagues carried out X-ray diffraction measurements on the 'rope' samples, in collaboration with John Fischer and co-workers at Pennsylvania State University. Well defined 2D lattice reflections were observed, confirming that the tubes had uniform diameters. A good fit to the experimental data was found by assuming a nanotube diameter of 1.38 nm, with an error of $\pm$ 0.02 nm. The van der Waals gap between the tubes was found to be 0.315 nm, similar to that in crystalline C_{60}. It was inferred from the XRD studies that the ropes were made up primarily of (10,10) armchair nanotubes. This was apparently confirmed by electron nanodiffraction measurements (2.63). Structural studies of nanotube rope samples are discussed further in the next chapter, and the possible growth mechanism put forward by Smalley and colleagues is summarised below.

2.6 Theories of nanotube growth

2.6.1 General comments

The growth mechanism of catalytic nanotubes was outlined in Section 2.3.2, and will not be discussed further here. This section is primarily concerned with the mechanisms by which nanotubes might grow in the arc. Unfortunately, arc-evaporation does not easily lend itself to mechanistic studies, and it has not been possible so far to determine the mechanism by direct experimentation. Therefore, any discussion of the growth mechanism in the arc must be mainly speculation.

To begin, it is worth considering the influence of tube structure on growth. Iijima pointed out in his 1991 *Nature* paper (2.7) that the growth of tubes with a helical structure would seem to be favoured, since such tubes have a repetitive step at the growing edge. This situation, illustrated in Fig. 2.20, is rather similar to the emergence of a screw dislocation from a crystal surface. Armchair and zig-zag nanotubes do not possess such a favourable growth structure and would require the repeated nucleation of a new ring of hexagons. This suggests that helical nanotubes should be much more commonly observed than armchair and zig-zag tubes, although at present there is insufficient experimental evidence to confirm this.

A further, very basic, question concerning the growth mechanism is whether growing tubes have closed or open ends. An early model of nanotube growth, put forward by Endo and Kroto, favoured a closed-end mechanism (2.64). They suggested that carbon atoms could be inserted into a closed fullerene surface at sites in the vicinity of pentagonal rings, followed by rearrangement to equilibrate strain, resulting in the continuous extension of an initial fullerene. In support of this idea, Endo and Kroto cited the demonstration by Ulmer and colleagues that C_{60} and C_{70} can apparently grow into larger fullerenes by the addition of small carbon fragments (2.65).

While the Endo–Kroto mechanism provides a plausible explanation for the growth of a single-layer nanotube, it has serious problems explaining multilayer growth. In their discussion of the model (2.64), Endo and Kroto suggest that multilayer growth may occur 'epitaxially'. If this is so, there seems no obvious reason why a second layer should not begin to grow almost as soon as the initial fullerene has formed; and as soon as the second layer becomes closed, any further extension of the inner tube would appear to be impossible. This is at odds with the observation that most tubes are multilayered over their entire length. The model also has difficulty explaining multiply compartmentalised structures such as those discussed in Chapter 3 (Fig. 3.20). For these reasons, the closed-end mechanism of Endo and Kroto has not been widely accepted.

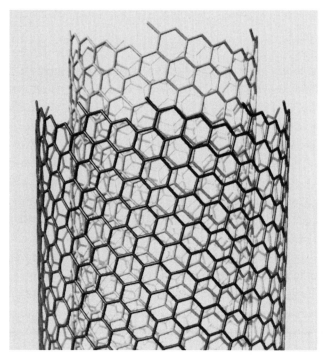

Fig. 2.20. Drawing of two concentric helical tubes showing the presence of steps at growing edges (2.9).

The conclusion that the growth mechanism must be open-ended is, in some ways, rather surprising. As Richard Smalley has stated, 'If we have learned anything about how carbon condenses since 1984–1985, it has been that open sheets will readily incorporate pentagons to eliminate dangling bonds' (2.66). The problem of tubes remaining open-ended under conditions favourable to closure is one with which a number of authors have grappled, and will now be discussed further.

2.6.2 *Why do tubes remain open during growth?*

Several authors, notably Smalley and colleagues, have suggested that the electric field in the arc may be important in keeping tubes open during growth (e.g. 2.14, 2.66). If correct, this would help to explain why nanotubes are never found in the soot which condenses on the walls of the arc-evaporation vessel, but only on the cathode. However, calculations indicated that field-induced lowering of the open tip energy was not sufficient to stabilise the open con-figuration except for unrealistically high fields (2.67, 2.68). Therefore, a refined model was developed in which adatom 'spot-welds' between layers help to stabilise the open tip conformation against closure (2.69). Support for this idea was provided by experiments on the closure of individual multiwalled nanotubes with and without the application of a bias voltage.

This model may help to explain nanotube growth in an arc, but it cannot account for tube growth under conditions where no strong electric fields are present (Sections 2.1.4, 2.1.5 and 2.2). This has led some authors to suggest that the interactions between adjacent concentric tubes may alone be sufficient to stabilise open tubes (2.70, 2.71). A detailed analysis of the interaction of two adjacent tubes was carried out by Jean-Christophe Charlier and colleagues using molecular dynamics simulations (2.70). They considered a (10,0) tube inside an (18,0) tube, and found that bridging bonds formed between the edges of the two tubes. At high temperatures (3000 K), the configuration of the lip–lip bonding structures was found to fluctuate continuously. It was suggested that this fluctuating structure would provide active sites for the adsorption and incorporation of new carbon atoms, thus enabling the tube to grow. A problem with this theory is that it cannot explain the growth of large-diameter single-walled tubes by heat-treatment of fullerene soot (Section 2.2). Overall, it appears that a complete explanation for the growth of open nanotubes is not available at present.

2.6.3 *Properties of the arc plasma*

Most models of nanotube growth discussed so far assume that the tubes nucleate and grow in the arc plasma. However, few authors have considered the physical state of the plasma and its role in nanotube formation. The most detailed discussion of this topic has been given by Eugene Gamaly, an expert on plasma physics, and Thomas Ebbesen (2.72, 2.73). This is a complex area, and only a brief summary is possible here.

Gamaly and Ebbesen begin by assuming that the nanotubes and nanoparticles form in the region of the arc next to the cathode surface. They then analyse the density and velocity of carbon vapours in this region, taking into account the temperature and the properties of the arc, in order to develop their model. They suggest that in the layer of carbon vapour adjacent to the cathode surface there will be two groups of carbon particles with different velocity distributions. This idea is central to their growth model. One group of carbon particles will have a Maxwellian, i.e. isotropic, velocity distribution corresponding to the temperature of the arc (~ 4000 K). The other group is composed of ions accelerated in the gap between the positive space charge and the cathode. The velocity of these carbon particles will be much greater than those of the thermal particles, and in this case the flux will be *directed* rather than isotropic. The process of nanotube (and nanoparticle) formation is considered to occur in a series of cycles, each of which comprises the following stages:

1 *Seed formation* At the beginning of the discharge process the velocity distribution of carbons in the vapour layer is predominantly Maxwellian, and this results in the formation of structures without any axis of symmetry such as nanoparticles. As the current becomes more directed, open structures begin to form which Gamaly and Ebbesen consider to be the seeds for nanotube growth.

2 *Tube growth during stable discharge* As the discharge stabilises, the current of carbon ions flows to the vapour layer in a direction perpendicular to the cathode surface. These directed carbons will contribute to the extension of single-walled and multi-walled nanotubes. Since the interaction of directed carbons with a solid surface will be much more intense than that of carbons in the vapour layer, the growth of elongated structures will be favoured over the formation of isotropic structures. However, the condensation of carbons from the vapour layer at the cathode surface will contribute to nanotube thickening.

3 *Termination and capping* Gamaly and Ebbesen point out that nanotubes are often observed to grow in bundles, and that in a given bundle all the tubes appear to begin and end growth at approximately the same time. This leads them to suggest that instabilities occur in the arc discharge which can lead to abrupt termination of nanotube growth. These instabilities might result from the erratic movement of the cathode spot around the cathode surface or from spontaneous interruption and re-striking of the arc. In such circumstances, carbons with a Maxwellian velocity distribution will again predominate, and the condensation of these carbons will tend to result in tube capping and the termination of growth.

2.6.4 An alternative model

A quite different theory of nanotube growth by arc-evaporation has been put forward by the present author and his colleagues (2.28). In this model, the nanotubes and nanoparticles do not grow in the arc plasma, but rather form on the cathode as a result of a solid-state transformation. Thus, nanotube growth is not a consequence of the electric field, but simply a result of the very rapid heating to high temperatures experienced by material deposited on the cathode during arcing. This idea was prompted by the observation that nanotubes can be produced by high-temperature heat treatment of fullerene soot, as described in Section 2.2, and envisages nanotube growth as a two-stage process, in which fullerene soot is an intermediate product. The model can be summarised as follows. In the initial stages of arc-evaporation, a fullerene soot-like material (plus fullerenes) would condense onto the cathode, and the condensed material would then experience extremely high temperatures as the arcing process continued, resulting in the formation firstly of single-layer, nanotube-like, structures and then of multilayer nanotubes.

In this two-stage model, the key process is the annealing of the fullerene soot. Thus, the soot which condenses on the reactor walls, which experiences relatively little annealing, is not transformed into tubes. On the other hand, the soot which condenses on the cathode must experience just the right amount of annealing: too much will result in sintering of tubes and nanoparticles into a solid mass. Therefore the model may enable us to understand the effect of variables such as electrode cooling and helium pressure on nanotube production. Water cooling seems to be essential in order to bring the cathode temperature down to the level required to avoid tube sintering. Similarly, the role of helium can be explained in terms of its effect on the temperature of the cathodic deposit. Since helium is an excellent thermal conductor, higher pressures would tend to reduce the electrode temperature, bringing it down to the region in which nanotube growth can occur without sintering.

If nanotube growth is simply due to high-temperature heating of fullerene soot, it might be asked why the experiments of de Heer and Ugarte produced primarily nanoparticles rather than nanotubes. Possibly, the answer may lie in the relatively slow rate at which the fullerene soot was heated in these experiments. In the arc, heating to very high temperatures occurs extremely rapidly, and this might be responsible for the one-dimensional growth. It hardly needs re-stating that this idea, like the other theories of nanotube growth discussed above, remains pure speculation at present.

2.6.5 Growth of single-walled nanotubes

We consider first the growth of single-walled nanotubes in an arc evaporator. This process raises at least as many questions as the growth of multiwalled tubes in the arc. Among the most obvious of these are: Why are only single-walled tubes observed? Why is there generally such a narrow distribution of tube diameters? What is the role of the metal? Why do the tubes frequently grow in the form of bundles? Once again, we have few definite answers to these questions.

One thing which seems clear is that the growth of single-walled tubes must be governed essentially by kinetics rather than thermodynamics, since very small tubes are expected to be less stable than larger ones (see Section 3.5). The absence of multiple layers is presumably also inhibited by kinetic factors. As far as the role of the metal is concerned, both Bethune and co-workers (2.52) and Iijima and Ichihashi (2.53) have suggested that individual metal atoms, or small clusters of atoms, might act as catalysts for growth in the vapour phase, by analogy with the way in which small metal particles catalyse the growth of multilayer nanotubes. The involvement of individual atoms, or well-defined small clusters, would help to explain the observed narrow size distributions. It is surprising, however, that catalytic particles are apparently never observed at the tips of the single-walled nanotubes. Even if the catalytic species were single atoms, it should be possible to detect these using high resolution electron microscopy or scanning transmission electron microscopy (STEM). Possibly, the catalytic atoms or particles become detached during the tube closure process. As noted above, Bethune and colleagues have shown that the addition of elements such as sulphur to the metal can strongly affect the distribution of tube diameters. Further work on this phenomenon might provide useful insights into the growth mechanism.

One of the few attempts to develop a detailed model of the growth of single-walled nanotubes has been made by Ching-Hwa Kiang and William Goddard (2.74). These workers suggest that planar polyyne rings might serve as nuclei for the formation of single-walled tubes. Such ring structures have been shown to be the dominant species in carbon vapours for C_{10}–C_{40}, while closed cage structures dominate for larger sizes (2.75). It has been postulated that carbon rings might be the precursors for fullerene formation, although this remains controversial (2.4, 2.5). Kiang and Goddard suggest that the starting materials for forming single-walled nanotubes are monocyclic carbon rings and gas phase cobalt carbide clusters, possibly charged. The cobalt carbide clusters act as catalysts for the addition of C_2 or other species to the rings. The authors suggest that the specific conformation of the initial ring would affect the structure of the resulting nanotube.

Following their synthesis of nanotube 'ropes' (2.6), Smalley and colleagues proposed a growth mechanism which has some similarities to that of Kiang and Goddard. This model is based on the assumption that all the tubes have the same (10,10) armchair structure. This structure uniquely allows the exposed hexagonal rings to be 'capped' with triple bonds, although these would be considerably strained from their preferred linear arrangement. The Smalley group then suggest that an individual nickel atom becomes chemisorbed to the edge and 'scoots' around the periphery (Fig. 2.21), assisting the arrangement of incoming carbon atoms into hexagonal rings. Any locally non-optimum structures, including pentagons, will be annealed out so the tube will continue to grow almost indefinitely.

A disadvantage of the scooter mechanism is that it only appears to work with armchair tubes, and recent work has cast some doubt on whether nanotube ropes are indeed mainly composed of such structures (see Section 3.6). Also, like the other mechanisms which have been proposed for the growth

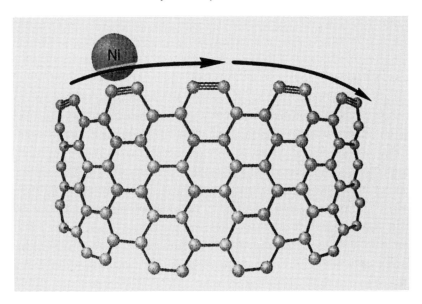

Fig. 2.21. Illustration of 'scooter' mechanism for growth of (10,10) armchair nanotubes (2.76).

of single-walled nanotubes, it is not supported by any direct experimental evidence.

2.7 Purification of multiwalled tubes

When produced by arc-evaporation, nanotubes are invariably accompanied by nanoparticles and other graphitic debris. Removing these side-products has proved difficult, and has impeded progress in testing the bulk properties of nanotubes. However, considerable progress has been made in this area recently, and some highly effective purification methods are now available.

The first successful technique for nanotube purification was developed by Thomas Ebbesen and co-workers (2.77). Following the demonstration that nanotube caps could be selectively attacked by oxidising gases (see Chapter 5), these workers realised that nanoparticles, with their defect-rich structures might be oxidised much more readily than the relatively perfect nanotubes. Therefore they subjected raw nanotube samples to a range of oxidising treatments, in the hope that the nanoparticles would be preferentially oxidised away. They found that a significant relative enrichment of nanotubes could be achieved in this way, but only at the expense of losing a major proportion of the original sample. Thus, in order to remove all the nanoparticles, it was necessary to oxidise more than 99% of the raw sample. When 95% of the original material was oxidised, about 10–20% of the remaining sample consisted of nanoparticles, while an oxidation of 85% resulted in no enrichment at all. These results suggest that the reactivities of nanotubes and nanoparticles towards oxidation are very similar, so that only a very narrow 'window' exists between the selective removal of nanoparticles and complete oxidation of the sample. Figure 2.22 (a) and (b) show low magnification micrographs of typical specimens before and after the oxidative treatment, illustrating that a very high degree of purification is achieved. However, the loss of over 99% of the original sample in this process is obviously a serious drawback.

An alternative approach to purifying multiwalled nanotubes was introduced in 1994 by a Japanese group (2.78). This technique made use of the fact that nanoparticles and other graphitic contaminants have relatively 'open' structures, and can therefore be more readily intercalated with a variety of materials than can closed nanotubes. By intercalating with copper chloride, and then reducing this to metallic copper, the Japanese group were able to preferentially oxidise the nanoparticles away, using copper as an oxidation catalyst. Since this has proved a popular method of nanotube purification, it is worth describing the procedure in more detail.

The first stage is to immerse the crude cathodic deposit in a molten $CuCl_2$–KCl mixture at a temperature of 400 °C, and leave for one week. The product of this treatment, which contains intercalated nanoparticles and graphitic fragments, is then washed in ion-exchanged water to remove excess $CuCl_2$ and KCl. In order to reduce the intercalated $CuCl_2$ to Cu metal, the washed product is slowly heated to 500 °C in a mixture of He and H_2, and held at this temperature for 1 hour. Finally, the material is oxidised in flowing air, heating at a rate of 10 °C/min to a temperature of 550 °C. Samples of cathodic soot which have been treated in this way consist almost entirely of nanotubes. A disadvantage of this method is that some nanotubes are inevitably lost in the oxidation stage, and the final material may be contaminated with residues of the intercalates. A similar purification technique, which involves intercalation with bromine followed by oxidation, has also been described (2.79).

A number of groups around the world have attempted to purify nanotube samples using methods such as centrifugation, filtration and chromatography. Several of these methods involve initially preparing colloidal suspensions of the nanotube-containing material using surface active agents. As an example, Jean-Marc Bonard and colleagues from the Ecole Polytechnique Fédérale de Lausanne in Switzerland employed the anionic surfactant sodium dodecyl sulphate (SDS) to achieve a stable suspension of nanotubes and nanoparticles in water (2.80). Initially, a filtration method was used to separate the nanotubes from the nanoparticles, but a more successful separation was achieved by

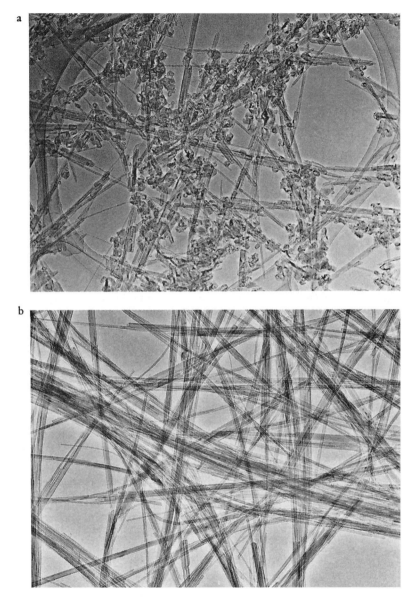

Fig. 2.22. Micrographs illustrating purification of multiwalled nanotube sample by oxidation (2.77).

simply allowing the nanotubes to flocculate, leaving the nanoparticles in suspension. The sediment could then be removed, and further flocculation procedures carried out. This not only enabled the nanoparticles to be removed, but also resulted in some degree of length separation of the tubes.

A different method of achieving size selection of nanotubes has been described by Duesberg and colleagues from the Max-Planck Institute in Stut-
tgart and Trinity College, Dublin (2.81). Again a dispersion of tubes and other
material was obtained with the aid of SDS. Separation was then carried out
using size exclusion chromatography (SEC). This technique is quite widely
used for the separation of biological macromolecules, and the authors demonstrated that it could successfully separate the nanotube samples into fractions
containing tubes of different lengths. One possible drawback of using surfac-
tants such as SDS in the purification of nanotubes is that traces of the
surfactant may remain in the final product, but Bonard and colleagues showed
that the level of SDS in the purified sample could be reduced to below 0.1% by
washing.

2.8 Purification of single-walled tubes

Methods have also been developed for the purification of single-walled
nanotubes, although this presents an even greater challenge than for multiwalled nanotubes. In addition to large amounts of amorphous carbon, the
nanotube-containing soot contains metal particles, often themselves coated
with carbon. Moreover, the harsh oxidation methods which can be used to
purify multiwalled nanotubes are too destructive to apply to single-walled
tubes.

A Japanese group has described a step-by-step process aimed at removing
the various contaminants in turn (2.82, 2.83). The first step involved refluxing
the raw soot in distilled water for 12 hours, followed by filtering and drying.
This treatment removed some of the graphitic particles and amorphous carbon. Fullerenes were then washed out using toluene in a Soxhlet apparatus.
Next, the soot was heated at 470 °C in air for 20 minutes, with the aim of
oxidising more of the amorphous carbon and removing the graphitic covering
from the metal particles. Finally, the remaining soot was treated with concentrated hydrochloric acid in order to dissolve the metal particles. Examination
of the final product by electron microscopy and X-ray diffraction indicated
that most contaminants had been removed, although some filled and empty
nanoparticles remained.

Smalley and co-workers have developed methods for purifying nanotube
rope samples using microfiltration. They first described a technique involving
the use of a cationic surfactant to suspend the nanotubes and accompanying
material in solution, and then trapping of the tubes on a membrane filter (2.84).
However, multiple filtration was required, with sample re-suspension after
each filtration, in order to achieve a significant level of purification, making the

procedure slow and inefficient. An improved method was described (2.85) in which ultrasonication was used to keep the material suspended during the filtration, thus enabling large amounts of sample to be filtered continuously. In this way, up to 150 mg of soot could be purified in 3–6 hours, with the resulting material containing more than 90% of SWNTs.

Single-walled tubes have also been purified using size exclusion chromatography. Duesberg *et al.* have described a method similar to that used for MWNTs (see previous section), which proved effective for SWNTs (2.86).

2.9 Alignment of nanotube samples

Many of the preparation methods described above produce samples in which the nanotubes are randomly oriented. Although the tubes are often grouped into bundles, the bundles themselves are not generally aligned with each other. In measuring nanotube properties, it would be very helpful to have samples in which *all* the tubes were aligned in a specific direction. Catalytic methods for preparing aligned tubes have already been described (Section 2.3.3), but techniques for aligning pre-synthesised samples of tubes are also needed. One of the first methods for doing so was described in early 1995 by a group from the Ecole Polytechnique Fédérale de Lausanne in Switzerland (2.87). They used a MWNT sample prepared by arc-evaporation, which was purified using centrifugation and filtration to remove the nanoparticles and other contaminating material. Thin films of the purified tubes were then deposited onto a plastic surface, and SEM images indicated that in this freshly deposited state the tubes were aligned perpendicular to the film. It was found that the tubes could be aligned parallel to the surface by the simple expedient of lightly rubbing with a Teflon or aluminium foil. The authors state that nanotube films can made 'arbitrarily large' by this method, and have used the films to carry out field emission experiments (see Section 4.8).

Another method of aligning nanotubes is to incorporate the tubes into a matrix and then extrude the matrix in some way, so that the tubes become aligned along the direction of flow. This has been achieved by a number of groups, and is discussed in Chapter 6 (p. 205).

2.10 Length control of carbon nanotubes

A technique for cutting individual single-walled nanotubes into controlled lengths was described by workers from Delft and Rice Universities in late 1997 (2.88). The nanotubes employed were prepared by the Smalley group's laser vaporisation technique, and were deposited onto single crystal gold surfaces for examination by scanning tunnelling microscopy. When a suitable nanotube had been identified, scanning was interrupted and the Pt/Ir tip was moved to a selected point on the tube. The feedback was then switched off and a voltage pulse applied between tip and sample for a specified period. When scanning was resumed, a break was visible in the nanotube if the cutting had been successful. It was demonstrated that individual tubes could be cut at up to four different positions. The critical factor in the cutting process was found to be the voltage rather than the current, the minimum voltage necessary to produce a cut being approximately 4 V.

Having broken individual nanotubes into short lengths, the authors were able to demonstrate that the electrical properties of the shortened tubes differed from those of the original nanotubes. These differences were attributed to quantum size effects.

As well as controlling the lengths of individual nanotubes, it is also possible to break down *bulk* samples of single-walled nanotubes into short lengths. This was demonstrated by Smalley's group in 1998 (2.89). The most effective way of producing samples of short tubes (which they named 'fullerene pipes') was found to be prolonged sonication of the nanotube material in a mixture of concentrated sulphuric and nitric acids. During this treatment, it appears that localised sonochemistry produces holes in the tube sides, which are then further attacked by the acids to leave the open 'pipes'. Smalley and colleagues showed that the pipes could be sorted into different length fractions by a method known as field-flow fractionation. They also derivitised the ends of the open nanotubes with various functional groups, and showed that gold particles could be attached to the functionalised tube ends. This work, together with some of the studies described in Chapter 5, could be seen as marking the beginning of a new organic chemistry based on carbon nanotubes.

2.11 Discussion

The Iijima–Ebbesen–Ajayan arc-evaporation method remains by far the best technique for the synthesis of high-quality nanotubes, but it suffers from a number of disadvantages. Firstly, it is labour intensive and requires some skill to achieve a satisfactory level of reproducibility. Secondly, the yield is rather low, since most of the evaporated carbon is deposited on the walls of the vessel rather than on the cathode, and the nanotubes are 'contaminated' with nanoparticles and other graphitic debris. Thirdly, it is a 'batch' rather than a continuous process, and it does not easily lend itself to scale-up. If carbon nanotubes are ever to be used commercially on a large scale it seems that some other preparation method will be needed. Progress in this direction has been hampered somewhat by a lack of understanding of the growth mechanism of tubes in the arc. Further studies devoted specifically to elucidating the mechanism of nanotube growth would be welcome.

There is a further serious weakness of the arc-evaporation method and of all current techniques for preparing multiwalled nanotubes: they produce a wide range of tube sizes and structures. This may not be a problem for some applications, but could be a drawback in areas where specific tube structures are needed, such as in nanoelectronics. Can one envisage a way in which tubes with defined structures could be prepared? Conceivably, this might be achieved by the creative use of catalysts. Of course, the catalytic synthesis of multiwalled nanotubes has been known for many years, but the quality of tubes produced in this way has been rather poor. However, it has been demonstrated recently that the adroit use of catalysts can lead to the formation of nanotubes that are both aligned and reasonably graphitic. This seems a promising direction for further research.

As far as single-walled nanotubes are concerned, these tend to be somewhat more homogenous than their multiwalled counterparts, at least in terms of their diameters. However, the methods currently used to synthesise single-walled tubes are rather more complicated than those for multiwalled tubes. The laser vaporisation technique developed by Smalley's group (Section 2.5.3) appears to produce the highest yield and best quality material, but the high powered lasers required for this method will obviously not be available in every laboratory. As with multiwalled tubes, the way forward may involve catalytic methods, and recent work in this area has been highly promising. Ultimately, we might hope that organic chemists could develop a complete synthesis of nanotubes. However, when one bears in mind that a total synthesis of C_{60} has not yet been achieved, this may be a distant prospect.

Given that the best quality nanotubes are currently made using methods which also produce significant amounts of contaminating material, it is important that methods for removing this material are available. Fortunately, this is an area where significant progress has been made recently, and there is now a range of methods for removing unwanted nanoparticles, microporous carbon and other contaminants from samples of both multiwalled and single-walled nanotubes. Procedures have also been developed for aligning tubes and cutting them into controlled lengths. These techniques will enable headway to be made in areas of nanotube research where, up to now, the lack of pure and well-defined samples has been a serious problem.

References

(2.1) J. R. Heath, S. C. O'Brien, R. F. Curl, H. W. Kroto and R. E. Smalley, 'Carbon condensation', *Comments Cond. Mat. Phys.*, **13**, 119 (1987).

(2.2) R. E. Smalley, 'Self-assembly of the fullerenes', *Acc. Chem. Res.*, **25**, 98 (1992).

(2.3) J. R. Heath, in *Fullerenes: synthesis, properties and chemistry of large carbon clusters*, ed. G. S. Hammond and V. J. Kuck, American Chemical Society, Washington DC, 1992, p. 1.

(2.4) R. F. Curl, 'On the formation of the fullerenes', *Philos. Trans. R. Soc. A*, **343**, 19 (1993).

(2.5) N. S. Goroff, 'Mechanism of fullerene formation', *Acc. Chem. Res.*, **29**, 77 (1996).

(2.6) A. Thess, R. Lee, P Nikolaev, H. Dai, P. Petit, J. Robert, C. Xu, Y. H. Lee, S. G. Kim, A. G. Rinzler, D. T. Colbert, G. E. Scuseria, D. Tománek, J. E. Fischer and R. E. Smalley, 'Crystalline ropes of metallic carbon nanotubes', *Science*, **273**, 483 (1996).

(2.7) S. Iijima, 'Helical microtubules of graphitic carbon', *Nature*, **354**, 56 (1991).

(2.8) T. W. Ebbesen and P. M. Ajayan, 'Large-scale synthesis of carbon nanotubes', *Nature*, **358**, 220 (1992).

(2.9) T. W. Ebbesen, 'Carbon nanotubes', *Ann. Rev. Mater. Sci.*, **24**, 235 (1994).

(2.10) T. W. Ebbesen in *Carbon nanotubes: preparation and properties*, ed. T. W. Ebbesen, CRC Press, Boca Raton, 1997, p. 139.

(2.11) Y. Saito, K. Nishikubo, K. Kawabata and T. Matsumoto, 'Carbon nanocapsules and single-layered nanotubes produced with platinum-group metals (Ru, Rh, Pd, Os, Ir, Pt) by arc-discharge', *J. Appl. Phys.* **80**, 3062 (1996).

(2.12) T. W. Ebbesen, H. Hiura, J. Fujita, Y. Ochiai, S. Matsui and K. Tanigaki, 'Patterns in the bulk growth of carbon nanotubes', *Chem. Phys. Lett.*, **209**, 83 (1993).

(2.13) G. H. Taylor, J. D. Fitzgerald, L. Pang and M. A. Wilson, 'Cathode deposits in fullerene formation – microstructural evidence for independent pathways of pyrolytic carbon and nanobody formation', *J. Cryst. Growth*, **135**, 157 (1994).

(2.14) D. T. Colbert, J. Zhang, S. M. McClure, P. Nikolaev, Z. Chen, J. H. Hafner, D. W. Owens, P. G. Kotula, C. B. Carter, J. H. Weaver, A. G. Rinzler and R. E. Smalley, 'Growth and sintering of fullerene nanotubes', *Science*, **266**, 1218 (1994).

(2.15) G. V. Coles, 'Occupational risks', *Nature*, **359**, 99, (1992).

(2.16) R. F. Service, 'Nanotubes: the next asbestos?', *Science*, **281**, 941 (1998).

(2.17) Z. Ja. Kosakovskaja, L. A. Chernozatonskii and E. A. Fedorov, 'Nanofilament carbon structures', *JETP Letters*, **56**, 26 (1992).

(2.18) L. A. Chernozatonskii, Z. Ja. Kosakovskaja, A. N. Kiselev and N. A. Kiselev, 'Carbon films of oriented multilayered nanotubes deposited on KBr and glass by electron beam evaporation', *Chem. Phys. Lett.*, **228**, 94 (1994).

(2.19) M. Ge and K. Sattler, 'Vapor-condensation generation and STM analysis of fullerene tubes', *Science*, **260**, 515 (1993).

(2.20) M. Ge and K. Sattler, 'Scanning tunnelling microscopy of single-shell nanotubes of carbon', *Appl. Phys. Lett.*, **65**, 2284 (1994).

(2.21) T. Guo, P. Nikolaev, A. G. Rinzler, D. Tomanek, D. T. Colbert and R. E. Smalley, 'Self assembly of tubular fullerenes', *J. Phys. Chem.*, **99**, 10,694 (1995).

(2.22) M. Endo, K. Takeuchi, S. Igarashi, K. Kobori, M. Shiraishi and H. W. Kroto,

'The production and structure of pyrolytic carbon nanotubes (PCNT's)', *J. Phys. Chem. Solids*, **54**, 1841 (1993).

(2.23) A. Sarkar, H. W. Kroto and M. Endo, 'Hemi-toroidal networks in pyrolytic carbon nanotubes', *Carbon*, **33**, 51 (1995).

(2.24) M. Endo, K. Takeuchi, K. Kobori, K. Takahashi, H. W. Kroto and A. Sarkar, 'Pyrolytic carbon nanotubes from vapor-grown carbon fibers', *Carbon*, **33**, 873 (1995).

(2.25) W. K. Hsu, J. P. Hare, M. Terrones, H. W. Kroto, D. R. M. Walton and P. J. F. Harris, 'Condensed phase nanotubes', *Nature*, **377**, 687 (1995).

(2.26) W. K. Hsu, M. Terrones, J. P. Hare, H. Terrones, H. W. Kroto and D. R. M. Walton, 'Electrolytic formation of carbon nanostructures', *Chem. Phys. Lett.*, **262**, 161 (1996).

(2.27) S. C. Tsang, P. J. F. Harris, J. B. Claridge and M. L. H. Green, 'A microporous carbon produced by arc-evaporation', *J. Chem. Soc., Chem. Commun.*, 1519 (1993).

(2.28) P. J. F. Harris, S. C. Tsang, J. B. Claridge and M. L. H. Green, 'High-resolution electron microscopy studies of a microporous carbon produced by arc-evaporation', *J. Chem. Soc., Faraday Trans.*, **90**, 2799 (1994).

(2.29) L. A. Bursill and L. N. Bourgeois, 'Image-analysis of a negatively curved graphitic sheet model for amorphous-carbon', *Mod. Phys. Lett. B*, **9**, 1461 (1995).

(2.30) S. J. Townsend, T. J. Lenosky, D. A. Muller, C. S. Nichols and V. Elser, 'Negatively curved graphitic sheet model of amorphous carbon', *Phys. Rev. Lett.*, **69**, 921 (1992).

(2.31) W. A. de Heer and D. Ugarte, 'Carbon onions produced by heat-treatment of carbon soot and their relation to the 217.5 nm interstellar absorption feature', *Chem. Phys. Lett.*, **207**, 480 (1993).

(2.32) D. Ugarte, 'High-temperature behaviour of "fullerene black"', *Carbon*, **32**, 1245 (1994).

(2.33) P. J. F. Harris and S. C. Tsang, 'High resolution electron microscopy studies of non-graphitizing carbons', *Philos. Mag. A*, **76**, 667 (1997).

(2.34) P. J. F. Harris, 'Structure of non-graphitising carbons', *International Materials Reviews*, **42**, 206 (1997).

(2.35) P. Schultzenberger and L. Schultzenberger, 'Sur quelques faits relatifs à l'histoire du carbone', *C. R. Acad. Sci., Paris*, **111**, 774 (1890).

(2.36) R. T. K. Baker and P. S. Harris, 'The formation of filamentous carbon', *Chem. Phys. Carbon*, **14**, 83 (1978).

(2.37) V. Ivanov, A. Fonseca, J. B. Nagy, A. Lucas, P. Lambin, D. Bernaerts and X. B. Zhang, 'Catalytic production and purification of nanotubules having fullerene-scale diameters', *Carbon*, **33**, 1727 (1995).

(2.38) M. Endo, 'Grow carbon fibres in the vapor phase', *Chemtech*, **18**, 568 (1988) (September issue).

(2.39) M. S. Dresselhaus, G. Dresselhaus, K. Sugihara, I. L. Spain and H. A. Goldberg, *Graphite fibers and filaments*, Springer-Verlag, Berlin, 1988.

(2.40) S. Amelinckx, X. B. Zhang, D. Bernaerts, X. F. Zhang, V. Ivanov and J. B. Nagy, 'A formation mechanism for catalytically grown helix-shaped graphite nanotubes', *Science*, **265**, 635 (1994).

(2.41) D. Bernaerts, X. B. Zhang, X. F. Zhang, S. Amelinckx, G. Van Tendeloo, J. Van Landuyt, V. Ivanov and J. B. Nagy, 'Electron microscopy study of coiled carbon tubules', *Philos. Mag. A*, **71**, 605 (1995).

(2.42) W. Z. Li, S. Xie, L. X. Qian, B. H. Chang, B. S. Zou, W. Y. Zhou, R. A. Zhao and G. Wang, 'Large-scale synthesis of aligned carbon nanotubes', *Science*, **274**, 1701 (1996).

(2.43) M. Terrones, N. Grobert, J. Olivares, J. P. Zhang, H. Terrones, K. Kordatos, W. K. Hsu, J. P. Hare, P. D. Townsend, K. Prassides, A. K. Cheetham, H. W. Kroto and D. R. M. Walton, 'Controlled production of aligned-nanotube bundles', *Nature*, **388**, 52 (1997).

(2.44) Z. F. Ren, Z. P. Huang, J. W. Xu, J. H. Wang, P. Bush, M. P. Siegal and P. N. Provencio, 'Synthesis of large arrays of well-aligned carbon nanotubes on glass', *Science*, **282**, 1105 (1998).

(2.45) P. J. F. Harris, 'Carbon nanotubes and other graphitic structures as contaminants on evaporated carbon films', *J. Microscopy*, **186**, 88 (1997).

(2.46) F. J. M. Rietmeijer, 'A poorly graphitized carbon contaminant in studies of extraterrestrial materials', *Meteoritics*, **20**, 43 (1985).

(2.47) S. Iijima, 'High resolution electron microscopy of some carbonaceous materials', *J. Microscopy*, **119**, 99 (1980).

(2.48) S. Iijima, 'Direct observation of the tetrahedral bonding in graphitized carbon black by high resolution electron microscopy', *J. Cryst. Growth*, **50**, 675 (1980).

(2.49) R. S. Ruoff, D. C. Lorents, B. Chan, R. Malhotra and S. Subramoney, 'Single crystal metals encapsulated in carbon nanoparticles', *Science*, **259**, 346 (1993).

(2.50) M. Tomita, Y. Saito and T. Hayashi, 'LaC_2 encapsulated in graphite nanoparticle', *Jap. J. Appl. Phys.*, **32**, L280 (1993).

(2.51) S. Seraphin, D. Zhou, J. Jiao, J. C. Withers and R. Loutfy, 'Yttrium carbide in nanotubes', *Nature*, **362**, 503 (1993).

(2.52) D. S. Bethune, C. H. Kiang, M. S. de Vries, G. Gorman, R. Savoy, J. Vasquez and R. Beyers, 'Cobalt-catalysed growth of carbon nanotubes with single-atomic-layer walls', *Nature*, **363**, 605 (1993).

(2.53) S. Iijima and T. Ichihashi, 'Single-shell carbon nanotubes of 1-nm diameter', *Nature*, **363**, 603 (1993).

(2.54) C. H. Kiang, W. A. Goddard III, R. Beyers and D. S. Bethune, 'Carbon nanotubes with single-layer walls', *Carbon*, **33**, 903 (1995).

(2.55) C. H. Kiang, P. H. M. van Loosdrecht, R. Beyers, J. R. Salem, D. S. Bethune, W. A. Goddard III, H. C. Dorn, P. Burbank and S. Stevenson, 'Novel structures from arc-vaporized carbon and metals: single-layer nanotubes and metallofullerenes', *Surf. Rev. Lett.*, **3**, 765 (1996).

(2.56) C. Journet, W. K. Maser, P. Bernier, A. Loiseau, M. Lamy de la Chapelle, S. Lefrant, P. Deniard, R. Lee and J. E. Fischer, 'Large-scale production of single-walled nanotubes by the electric-arc technique', *Nature*, **388**, 756 (1997).

(2.57) S. Subramoney, R. S. Ruoff, D. C. Lorents and R. Malhotra, 'Radial single-layer nanotubes', *Nature*, **366**, 637 (1993).

(2.58) Y. Saito, M. Okuda, M. Tomita and T. Hayashi, 'Extrusion of single-wall carbon nanotubes via formation of small particles condensed near an arc evaporation source', *Chem. Phys. Lett.*, **236**, 419 (1995).

(2.59) D. Zhou, S. Seraphin and S. Wang, 'Single-walled carbon nanotubes growing radially from YC_2 particles', *Appl. Phys. Lett.*, **65**, 1593 (1994).

(2.60) R. T. K. Baker and R. J. Waite, 'Formation of carbonaceous deposits from catalysed decomposition of acetylene', *J. Catalysis*, **37**, 101 (1975).

(2.61) H. Dai, A. G. Rinzler, P. Nikolaev, A. Thess, D. T. Colbert and R. E. Smalley, 'Single-wall nanotubes produced by metal-catalyzed disproportionation of carbon monoxide', *Chem. Phys. Lett.*, **260**, 471 (1996).

(2.62) T. Guo, P. Nikolaev, A. Thess, D. T. Colbert and R. E. Smalley, 'Catalytic growth of single-walled nanotubes by laser vaporization', *Chem. Phys. Lett.*, **243**, 49 (1995).

(2.63) J. M. Cowley, P. Nikolaev, A. Thess and R. E. Smalley, 'Electron nano-diffraction study of carbon single-walled nanotube ropes', *Chem. Phys. Lett.*, **265**, 379 (1997).

(2.64) M. Endo and H. W. Kroto, 'Formation of carbon nanofibers', *J. Phys. Chem.*, **96**, 6941 (1992).

(2.65) G. Ulmer, E. E. B. Campbell, R. Kuhnle, H.-G. Busmann and I. V. Hertel, 'Laser mass spectroscopic investigations of purified, laboratory-produced C_{60}/C_{70}', *Chem. Phys. Lett.*, **182**, 114 (1991).

(2.66) R. E. Smalley, 'From dopyballs to nanowires', *Mater. Sci. Eng. B*, **19**, 1 (1993).

(2.67) A. Maiti, C. J. Brabec, C. M. Roland and J. Bernholc, 'Growth energetics of carbon nanotubes', *Phys. Rev. Lett.*, **73**, 2468 (1994).

(2.68) L. Lou, P. Nordlander and R. E. Smalley, 'Fullerene nanotubes in electric fields', *Phys. Rev. B*, **52**, 1429 (1995).

(2.69) D. T. Colbert and R. E. Smalley, 'Electric effects in nanotube growth', *Carbon*, **33**, 921 (1995).

(2.70) J.-C. Charlier, A. De Vita, X. Blase and R. Car, 'Microscopic growth mechanisms for carbon nanotubes', *Science*, **275**, 646 (1997).

(2.71) Y.-K. Kwon, Y. H. Lee, S.-G. Kim, P. Jund, D. Tomanek and R. E. Smalley, 'Morphology and stability of growing multiwall carbon nanotubes', *Phys. Rev. Lett.*, **79**, 2065 (1997).

(2.72) E. G. Gamaly and T. W. Ebbesen, 'Mechanism of carbon nanotube formation in the arc discharge', *Phys. Rev. B*, **52**, 2083 (1995).

(2.73) E. G. Gamaly in *Carbon nanotubes: preparation and properties*, ed. T. W. Ebbesen, CRC Press, Boca Raton, 1997, p. 163.

(2.74) C. H. Kiang and W. A. Goddard III, 'Polyyne ring nucleus growth model for single-layer carbon nanotubes', *Phys. Rev. Lett.*, **76**, 2515 (1996).

(2.75) J. M. Hunter, J. L. Fye, E. J. Roskamp and M. F. Jarrold, 'Annealing carbon cluster ions – a mechanism for fullerene synthesis', *J. Phys. Chem.*, **98**, 1810 (1994).

(2.76) P. Ball, 'The perfect nanotube', *Nature*, **382**, 207 (1996).

(2.77) T. W. Ebbesen, P. M. Ajayan, H. Hiura and K. Tanigaki, 'Purification of carbon nanotubes', *Nature*, **367**, 519 (1994).

(2.78) F. Ikazaki, S. Ohshima, K. Uchida, Y. Kuriki, H. Hayakawa, M. Yumura, K. Takahashi and K. Tojima, 'Chemical purification of carbon nanotubes by the use of graphite intercalation compounds', *Carbon*, **32**, 1539 (1994).

(2.79) Y. J. Chen, M. L. H. Green, J. L. Griffin, J. Hammer, R. M. Lago and S. C. Tsang, 'Purification and opening of carbon nanotubes via bromination', *Advanced Materials*, **8**, 1012 (1996).

(2.80) J.-M. Bonard, T. Stora, J.-P. Salvetat, F. Maier, T. Stöckli, C. Duschl, L. Forró, W. A. de Heer and A. Châtelain, 'Purification and size-selection of carbon nanotubes', *Advanced Materials*, **9**, 827 (1997).

(2.81) G. S. Duesberg, M. Burghard, J. Muster, G. Philipp and S. Roth, 'Separation of carbon nanotubes by size exclusion chromatography', *J. Chem. Soc., Chem. Commun.*, 435 (1998).

(2.82) K. Tohji, T. Goto, H. Takahashi, Y. Shinoda, N. Shimizu, B. Jeyadevan, I. Matsuoka, Y. Saito, A. Kasuya, T. Ohsuna, H. Hiraga and Y. Nishina, 'Purifying single-walled nanotubes', *Nature*, **383**, 679 (1996).

(2.83) K. Tohji, H. Takahashi, Y. Shinoda, N. Shimizu, B. Jeyadevan, I. Matsuoka, Y. Saito, A. Kasuya, S. Ito and Y. Nishina, 'Purification procedure for single-walled nanotubes', *J. Phys. Chem. B*, **101**, 1974 (1997).

(2.84) S. Bandow, A. M. Rao, K. A. Williams, A. Thess, R. E. Smalley and P. C. Eklund, 'Purification of single-wall carbon nanotubes by microfiltration', *J. Phys. Chem. B*, **101**, 8839 (1997).

(2.85) K. B. Shelimov, R. O. Esenaliev, A. G. Rinzler, C. B. Huffman and R. E. Smalley, 'Purification of single-wall nanotubes by ultrasonically assisted filtration', *Chem. Phys. Lett.*, **282**, 429 (1998).

(2.86) G. S. Duesberg, J. Muster, V. Krstic, M. Burghard and S. Roth, 'Chromatographic size separation of single-wall carbon nanotubes', *Appl. Phys. A*, **67**, 117 (1998).

(2.87) W. A. de Heer, W. S. Bacsa, A. Châtelain, T. Gerfin, R. Humphrey-Baker, L. Forro and D. Ugarte, 'Aligned carbon nanotube films: production and optical and electronic properties', *Science*, **268**, 845 (1995).

(2.88) L. C. Venema, J. W. G. Wildoer, H. L. J. T. Tuinstra, C. Dekker, A. G. Rinzler and R. E. Smalley, 'Length control of individual carbon nanotubes by nanostructuring with a scanning tunnelling microscope', *Appl. Phys. Lett.*, **71**, 2629 (1997).

(2.89) J. Liu, A. G. Rinzler, H. Dai, J. H. Hafner, R. K. Bradley, P. J. Boul, A. Lu, T. Iverson, K. Shelimov, C. B. Huffman, F. Rodriguez-Macias, Y.-S. Shon, T. R. Lee, D. T. Colbert and R. E. Smalley, 'Fullerene pipes', *Science*, **280**, 1253 (1998).

3

Structure

Particularly, I have urged them to learn what they can of chemistry, for I feel that chemistry is basic structure, *ergo* architecture.
Buckminster Fuller, *The Comprehensive Man*

Classifying the structures of carbon nanotubes poses an interesting challenge. Although essentially crystalline, these tubular structures cannot be treated in terms of the conventional crystallography of three-dimensional solids. Theoretical methods have been developed for analysing cylindrical arrays in biology but these are insufficient for a full analysis of nanotube structure. New methods are therefore needed. Fortunately, several groups have developed such methods, and a complete framework for analysing the structures and symmetries of cylindrical nanotubes is now available. These methods have been essential in determining the electronic and vibrational properties of nanotubes, as discussed in the next chapter. Theoretical discussions have also been given of the layer structure of multiwalled tubes, of helically coiled tubes and of other aspects of nanotube structure such as elbow connections.

Experimental studies of nanotube structure have mainly been carried out using high resolution electron microscopy, and many beautiful images revealing the intricate structure of multiwalled nanotube caps, internal compartments and so on have been published. Atomic force microscopy (AFM) and scanning tunnelling microscopy (STM) have proved more difficult to apply to nanotubes but some useful images, particularly of single-walled tubes, have been achieved recently. As a result, we now have a reasonable understanding of the main structural features of both multiwalled and single-walled nanotubes.

This chapter begins with a discussion of tubular biological structures and of bonding in graphite and fullerenes. Theoretical models of carbon nanotube structure are then summarised. Following a brief section on the stability of carbon nanotubes, experimental work on the structures of nanotubes and nanotube-related particles is discussed. Multiwalled nanotubes are considered first, followed by single-walled tubes, and associated structures such as nanohoops. Finally, structural aspects of carbon nanoparticles and nanocones are discussed.

3.1 Classification of tubular biological structures

Tubular structures assembled from a regular packing of protein monomers occur quite widely in biology. Microtubules, which perform a variety of functions in higher organisms, have this form, as do the flagella of some bacteria. But the closest parallel with carbon nanotubes is found in the protein coats of viruses. Just as C_{60} and other symmetrical fullerenes have their counterparts in icosahedral viruses, certain bacteriophages (bacteriolytic viruses) have structures remarkably similar to those of carbon nanotubes; an example is shown in Fig. 3.1. This shows the capsid or head of bacteriophage φCbK, which consists of a tubule with a hexagonal arrangement of protein units. The structure is capped with hemispherical fullerene-like domes.

A detailed discussion of the close packing of spheres in cylindrical structures was carried out by Ralph Erickson of the University of Pennsylvania (3.2). He began by considering symmetrical patterns of *points* on a cylinder. A simple example is the square lattice shown in Fig. 3.2. In this case, all points lie on what Erickson called a single generative helix. Thus, a single helix can be traced around the structure by joining points 0, 1, 2, . . . and so on, and will include all points. Note that the points making up the generative helix are not necessarily adjacent. Symmetrical patterns can also be constructed such that the points lie on two or more generative helices, in which case two or more points can be found on the same circular circumference of the cylinder. Erickson used the term 'jugacy' to describe the number of generative helices needed to construct a cylindrical array of points: patterns with 2, 3 or k generative helices were said to be bijugate, trijugate or k-jugate. Erickson specified the cylindrical arrays both by their jugacy and by citing two or more sets of 'parastichies' running through the pattern. Parastichies can be defined as helices made up of adjacent points on the cylinder. When a pattern is specified by two sets of parastichies, these are referred to as x-parastichies and y-parastichies, where x and y are integers with $x < y$. The meaning of x and y can be seen with reference to Fig. 3.2, in which the 3-, 5- and 8-parastichies are indicated by solid lines. Thus, one of the 5-parastichies, for example, joins points 0, 5, 10, 15, . . . while another joins points 1, 6, 11, 16, ... Clearly there are five 5-parastichies, eight 8-parastichies, and so on. In addition to x and y, other parameters are needed to describe completely a cylindrical lattice of points, but these need not be discussed here.

a

b

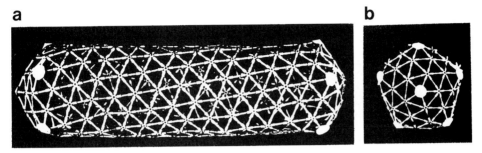

Fig. 3.1. Geodesic model of bacteriophage φCbK capsid (3.1).

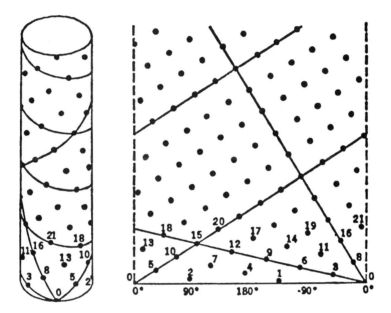

Fig. 3.2. A square lattice of points on a cylindrical surface, and unrolled into a plane. The solid lines indicate 3-, 5- and 8-parastichies (3.2).

In the next stage of his analysis, Erickson considered the uniform packing of spheres on a cylinder. This is a special case of the regular distribution of points on a cylinder, with restrictions which depend on the nature of the packing. Thus, for rhombic packing the points must be equidistant along two parastichies, while for hexagonal close packing the points must be equidistant along three parastichies. It is important to note that Erickson only considered *contact* parastichies when discussing packings of spheres, so that in the case of rhombic packing there are *only* two possible parastichies, x and y, and for hexagonal close packing only three, x, y and $x + y$. Figure 3.3 shows a number

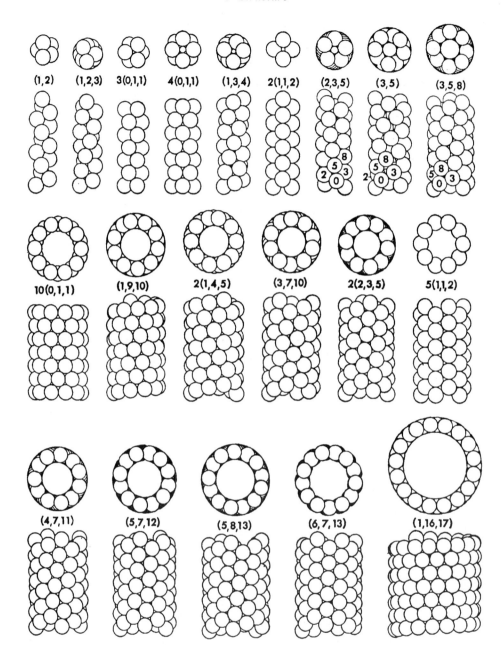

Fig. 3.3. Tubular arrangements of spheres, in side view and top view, with jugacy and parastichy numbers (3.2).

of tubular arrangements of spheres, in side view and top view, with the parameters defining each pattern. The numbers in brackets are the parastichy numbers, while a number appearing outside the brackets is the jugacy, k, or number of generative helices. It should be noted that the jugacy is equivalent to the frequency of rotational symmetry, such that a k-jugate pattern is symmetrical on rotation through $360/k$ degrees.

Erickson derived equations for parameters such as the angle of rotation of successive spheres on the cylinder and the angle of incidence of the generative helix in terms of x and y. He could then analyse a number of biological structures by firstly determining their jugacy and then measuring the angles of inclination of one or more sets of contact parastichies. Although it is not possible to apply all aspects of Erickson's analysis to carbon nanotubes, his discussion provides a valuable insight into the nature of helical tubular structures.

3.2 Bonding in carbon materials

A free carbon atom has the electronic structure $(1s)^2(2s)^2(2p)^2$. In order to form covalent bonds, one of the 2s electrons is promoted to 2p, and the orbitals are then hybridised in one of three possible ways. In graphite, one of the 2s electrons hybridises with two of the 2p's to give three sp^2 orbitals at $120°$ to each other in a plane, with the remaining orbital having a p_z configuration, at $90°$ to this plane. The sp^2 orbitals form the strong σ bonds between carbon atoms in the graphite planes, while the p_z, or π, orbitals provide the weak van der Waals bonds between the planes. The overlap of π orbitals on adjacent atoms in a given plane provides the electron bond network which gives graphite its relatively high electrical conductivity. In naturally occurring or high-quality synthetic graphite, the stacking sequence of the layers is generally ABAB, with an interlayer {0002} spacing of approximately 0.334 nm, as shown in Fig. 3.4. This structure is often known as Bernal graphite after John D. Bernal who first proposed it in 1924. The unit cell contains four atoms, and the space group is $P6_3/mmc$ (D_{6h}^4). In less perfect graphites, the interplanar spacing is found to be significantly larger than the value for single crystal graphite (typically ~ 0.344 nm), and the layer planes are randomly rotated with respect to each other about the c axis. Such graphites are termed turbostratic.

In diamond, each carbon atom is joined to four neighbours in a tetrahedral structure. The bonding here is sp^3 and results from the mixing of one 2s and three 2p orbitals. Diamond is less stable than graphite, and is converted to graphite at a temperature of 1700 °C at normal pressures. Disordered carbons

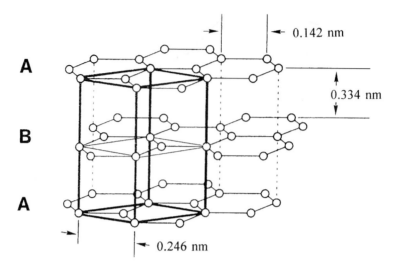

Fig. 3.4. The structure of hexagonal (Bernal) graphite, showing the unit cell.

containing sp^3-bonded atoms are also rapidly transformed into graphitic carbon at high temperatures.

In the C$_{60}$ molecule, shown in Fig. 1.1, the carbon atoms are bonded in an icosahedral structure made up of 20 hexagons and 12 pentagons. Each of the carbon atoms in C$_{60}$ is joined to three neighbours, so the bonding is essentially sp^2, although there may be a small amount of sp^3 character due to the curvature. Note that all 60 carbon atoms are identical, so that the strain is evenly distributed over the molecule. The bonding in carbon nanoparticles and nanotubes is also primarily sp^2, although once again there may be some sp^3 character in regions of high curvature.

3.3 The structure of carbon nanotubes: theoretical discussion

3.3.1 Vector notation for carbon nanotubes

As mentioned in Chapter 1, there are two possible high symmetry structures for nanotubes, known as 'zig-zag' and 'armchair'. These are illustrated in Fig. 1.2. In practice, it is believed that most nanotubes do not have these highly symmetric forms but have structures in which the hexagons are arranged helically around the tube axis, as in Fig. 3.5. These structures are generally known as chiral, since they can exist in two mirror-related forms.

The simplest way of specifying the structure of an individual tube is in terms of a vector, which we label **C**, joining two equivalent points on the

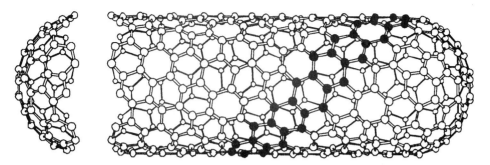

Fig. 3.5. Drawing of a chiral nanotube (adapted from Ref. (3.3)).

original graphene lattice. The cylinder is produced by rolling up the sheet such that the two end-points of the vector are superimposed. Because of the symmetry of the honeycomb lattice, many of the cylinders produced in this way will be equivalent, but there is an 'irreducible wedge' comprising one twelfth of the graphene lattice, within which unique tube structures are defined. Figure 3.6 shows a small part of this irreducible wedge, with points on the lattice labelled according to the notation of Dresselhaus *et al.* (3.4, 3.5). Each pair of integers (n,m) represents a possible tube structure. Thus the vector $\mathbf{C}$ can be expressed as

$$\mathbf{C} = n\mathbf{a}_1 + m\mathbf{a}_2$$

where $\mathbf{a}_1$ and $\mathbf{a}_2$ are the unit cell base vectors of the graphene sheet, and $n \geq m$. It can be seen from Fig. 3.6 that $m = 0$ for all zig-zag tubes, while $n = m$ for all armchair tubes. All other tubes are chiral. In the case of the two 'archetypal' nanotubes which can be capped by one half of a C_{60} molecule, the zig-zag tube is represented by the integers (9,0) while the armchair tube is denoted by (5,5). Since $|\mathbf{a}_1| = |\mathbf{a}_2| = 0.246$ nm, the magnitude of $\mathbf{C}$ in nanometres is $0.246\sqrt{(n^2 + nm + m^2)}$, and the diameter d_t is given by

$$d_t = 0.246\sqrt{(n^2 + nm + m^2)}/\pi.$$

The chiral angle, θ, is given by

$$\theta = \sin^{-1}\frac{\sqrt{3}m}{2\sqrt{(n^2 + nm + m^2)}}. \tag{3.1}$$

It might also be noted that the jugacy (see Section 3.2) of a tube is equal to m.

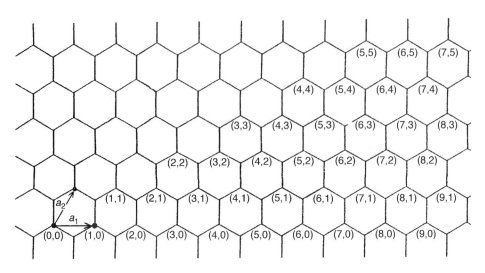

Fig. 3.6. Graphene layer with atoms labelled using (*n,m*) notation. Unit vectors of the 2D lattice are also shown.

3.3.2 *Unit cells of nanotubes*

If we think of a nanotube as a 'one-dimensional crystal', we can define a translational unit cell along the tube axis. For all nanotubes, the translational unit cell has the form of a cylinder. Considering again the two archetypal tubes which can be capped by one half of a C_{60} molecule, the 'unrolled' cylindrical unit cells for both of these are shown in Fig. 3.7. For the armchair tube, the width of the cell is equal to the magnitude of **a**, the unit vector of the original 2D graphite lattice, while for the zig-zag tube the width of the cell is $\sqrt{3}\mathbf{a}$. Larger diameter armchair and zig-zag nanotubes have unit cells which are simply longer versions of these. For chiral nanotubes, the lower symmetry results in larger unit cells. A simple method of constructing these cells has been described by Jishi, Dresselhaus and colleagues (3.4–3.7). This involves drawing a straight line through the origin O of the irreducible wedge normal to **C**, and extending this line until it passes exactly through an equivalent lattice point. This is illustrated in Fig. 3.8 for the case of a (6,3) nanotube. The length of the unit cell in the tube axis direction is the magnitude of the vector **T**. Expressions can be derived for this in terms of *C*, the magnitude of **C**, and the highest common divisor of *n* and *m*, which we denote d_{H} (3.5, 3.6). If $n - m \neq 3rd_{\mathrm{H}}$, where *r* is some integer, then

$$T = \sqrt{3}C/d_{\mathrm{H}},$$

while if $n - m = 3rd_{\mathrm{H}}$, then

(a)

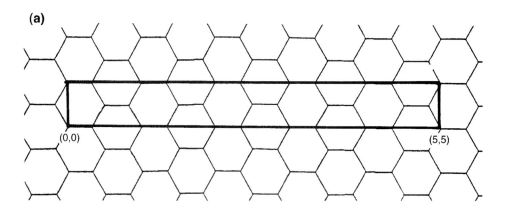

(b)

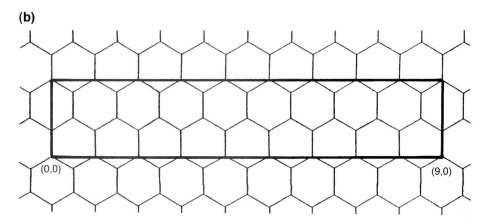

Fig. 3.7. Unit cells for (a) (5,5) armchair nanotube and (b) (9,0) zig-zag nanotube.

$$T = \sqrt{3}C/3d_{\mathrm{H}}.$$

It can also be shown that the number of carbon atoms per unit cell of a tube specified by (n,m) is $2N$ such that

$$N = 2(n^2 + m^2 + nm)/d_{\mathrm{H}} \qquad \text{if } n - m \neq 3rd_{\mathrm{H}}$$

and

$$N = 2(n^2 + m^2 + nm)/3d_{\mathrm{H}} \qquad \text{if } n - m = 3rd_{\mathrm{H}}.$$

These simple expressions enable the diameters and unit cell parameters of nanotubes to be readily calculated. For nanotubes in the diameter range which is typically observed experimentally, i.e. $\sim$ 2–30 nm, the unit cells can be very large. For example, the tube denoted (80, 67), which has a diameter of approxi-

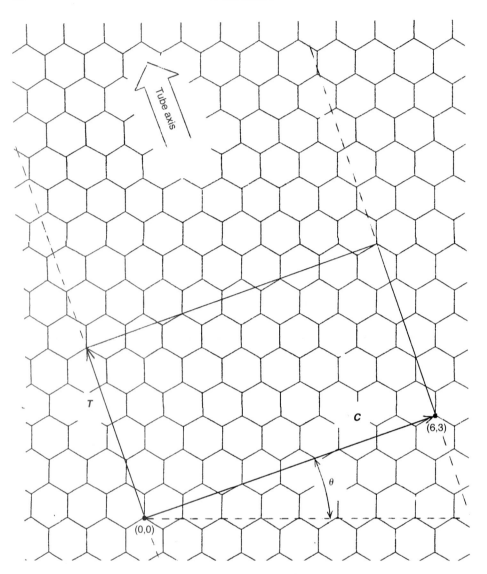

Fig. 3.8. Construction of the unit cell for a (6,3) nanotube.

mately 10 nm, has a unit cell 54.3 nm in length containing 64 996 atoms. These large unit cells can present problems in calculating the electronic and vibrational properties of nanotubes. For this reason, White, Mintmire and colleagues have proposed an alternative method of generating tube structures, which makes use of helical operators rather than a translational unit cell. This will not be detailed here, but has been described in a number of papers and reviews (e.g. 3.8, 3.9).

3.3.3 Multiwalled nanotubes

Nanotubes produced by the conventional arc-evaporation method invariably contain at least two concentric layers, so it is important to address the question of the structural relationship between successive cylinders. This has been discussed by Zhang and colleagues (3.10) and by Reznik *et al.* (3.11), who have reached broadly similar conclusions. In the discussion which follows, it is assumed that the tubes are concentric rather than 'scroll-like' in structure. Experimental studies indicate that this assumption is probably correct for most real tubes (see Section 3.5.2).

If the concentric graphene tubes are separated by a distance of approximately 0.334 nm, then successive tubes should differ in circumference by $(2\pi \times 0.334)$ nm ≈ 2.1 nm. It can readily be seen that this is not possible for zig-zag tubes, since 2.1 nm is not a precise multiple of 0.246 nm, the width of one hexagon. The closest approximation to the 'correct' separation is obtained if two successive cylinders differ by 9 rows of hexagons, which produces an inter-tube distance of 0.352 nm. A schematic section through a three-layer zig-zag tube, reproduced from Zhang *et al.* (3.10), is shown in Fig. 3.9. Here, the bold lines indicate the 9 and 18 extra rows of atoms that have been added to the centre and outer tubes respectively. The inclusion of these extra rows is similar to the introduction of a Shockley partial dislocation. It is clear from Fig. 3.9 that, for the most part, the ABAB stacking of perfect graphite is not present in concentric zig-zag tubes. However, short regions exist half-way between each 'dislocation' in which there is a good approximation to ABAB stacking.

In the case of armchair tubes, multiwalled structures can be assembled in which the ABAB arrangement is maintained and the interlayer distance is 0.34 nm. This is because 2.1 nm is close to 5×0.426 nm, the length of the repeat unit from which armchair tubes are constructed. For chiral nanotubes, the situation is complicated, but in general it is not possible to have two tubes with exactly the same chiral angle separated by the graphite interplanar distance. Overall it seems unlikely that the ABAB stacking of single-crystal graphite will be present in cylindrical carbon nanotubes, except possibly in small areas.

The energetics of multiwalled carbon tubes have been considered by Charlier and Michenaud (3.12). They found that the energy gained by adding a new cylindrical layer to a central one was of the same order as the one in graphite bilayering. The optimum interlayer distance between an inner (5,5) nanotube and an outer (10,10) tube was found to be 0.339 nm. This is somewhat smaller than the 0.344 nm {0002} spacing found in turbostratic graphite. Experimental measurements show that the interlayer spacings in nanotubes can vary quite considerably, but are typically around 3.4 nm (Section 3.5.2). Charlier and

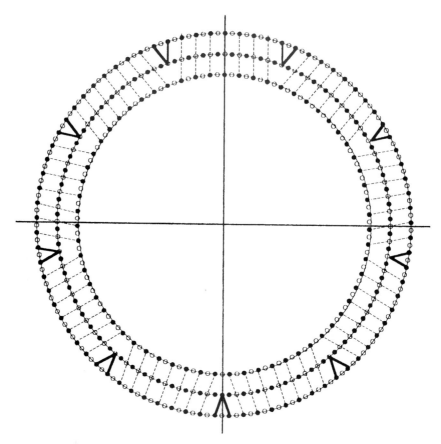

Fig. 3.9. Schematic illustration of three-layer nanotube showing how 'interfacial dislocations' (bold lines) can be introduced to accommodate strains. Full circles represent atoms in plane of paper, open circles atoms out of plane of paper (3.10).

Michenaud also estimated the translational and rotational energy barriers for the two coaxial tubes, and found the values of 0.23 mev/atom and 0.52 mev/atom. These low values suggest that significant translational and rotational mobility could be present in multilayered nanotubes at room temperature, although in reality the presence of caps and defects in the cylindrical structures would limit this mobility.

3.3.4 Theory of nanotube capping

It has been established that there are a very large number of possible cylindrical graphene structures. Experimentally, we know that nanotubes are almost invariably closed at both ends, so it is important to ask whether all of the possible cylinders are capable of being capped. This question has been considered by Fujita, Dresselhaus and colleagues (3.5, 3.13, 3.14) who have shown that in fact all nanotubes larger than the archetypal (5,5) and (9,0) tubes can be capped, and that the number of possible caps increases rapidly with increasing diameter. The following summary draws largely on their work.

Like fullerenes, all capped nanotubes must obey Euler's law. This states that a hexagonal lattice of any size or shape can only form a closed structure by the inclusion of precisely 12 pentagons. Therefore, any nanotube cap must contain 6 pentagons, and considerations of strain dictate that these pentagons must be isolated from each other (neglecting, for the moment, caps containing heptagons). As noted above, the smallest tubes which can be capped with *isolated* pentagons are the two archetypal tubes shown in Fig. 1.2, and for each of these there is only one possible cap, corresponding to the C_{60} molecule divided in two different ways. Fujita *et al.* have calculated the number of possible caps for nanotubes larger than these, using a method based on 'projection mapping' (3.13). This method involves constructing a map on a honeycomb network, which can be folded to form a given fullerene or nanotube. The pentagons are introduced by removing a 60° triangular segment of lattice, resulting in the formation of a conical defect known as a 60° positive wedge disclination, as shown in Fig. 3.10(a). Following Fujita *et al.* we firstly consider the projection mapping of icosahedral fullerenes, and then show how it can be extended to nanotubes. An icosahedral fullerene can be fully specified by the vector which connects two adjacent pentagons. As an illustration, consider the icosahedral fullerene C_{140}. In this case the defining vector, which we designate (m_f, n_f), is (2,1), as shown in Fig. 3.10(b). The complete projection map for C_{140} is shown in Fig. 3.11; in this case the defects form a regular triangular array. The fullerene is formed by removing the non-shaded part of the lattice and superimposing rings with the same numbers.

Now consider a nanotube capped at each end with one half of a C_{140} molecule. This can be mapped by simply extending the two lines AC and BD. The resulting tube is chiral, with the vector (10,5). Fujita *et al.* showed that a general icosahedral fullerene designated by the indices (m_f, n_f), when divided in half in a direction perpendicular to one of the five-fold axes, will cap a nanotube with the indices $(5m_f, 5n_f)$ (3.13). Thus, the series of so-called 'magic number' icosahedral fullerenes, C_{60}, C_{240}, C_{540}, . . ., which have the indices (1,1), (2,2), (3,3), . . ., can be bisected to cap the series of armchair tubes with the vectors (5,5), (10,10), (15,15) and so on. Similarly, when bisected in a direction perpendicular to one of the three-fold axes, these fullerenes will cap the tubes (9,0), (18,0), (27,0) etc.

(a)

(b)

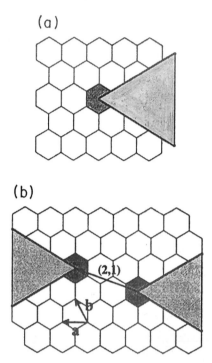

Fig. 3.10. (a) Creation of a pentagonal defect in a hexagonal lattice by removal of the shaded area. (b) The vector (m_f,n_f) connecting two pentagonal defects which specify the icosahedral fullerene C_{140} (3.13).

As already indicated, all nanotubes larger than the (5,5) and (9,0) tubes (with one exception) can be capped in more than one way. This is illustrated in Fig. 3.12 which shows two different ways of capping the chiral nanotube defined by the vector (7,5). In fact, Dresselhaus and colleagues have shown that there are 13 possible caps for this tube.

Experimental studies of nanotube caps, described in Section 3.5.7, show that they frequently have conical shapes, so it is worth considering the possible cone angles which are formed by the introduction of pentagonal rings into a hexagonal network. A cone is formed by the introduction of fewer pentagons than the six needed to form a cylinder. It can be shown quite easily that the cone angle, α, is given by:

$$\sin(\alpha/2) = 1 - (n_p/6)$$

where n_p is the number of pentagons in the cone. Values for the opening angles of cones containing one to five pentagonal rings are given in Table 3.1.

Table 3.1. *Cone angles for graphitic cones with various numbers of pentagons*

Number of pentagons	Cone angles in degrees
1	112.9
2	83.6
3	60.0
4	38.9
5	19.2
6	0.0

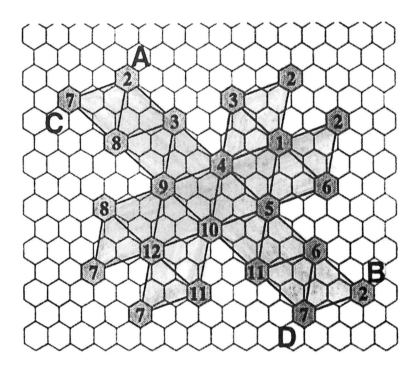

Fig. 3.11. Projection map for C_{140} (3.13).

3.3.5 Symmetry classification of nanotubes

We now consider the symmetry classification of carbon nanotubes, once again following the work of Mildred Dresselhaus and co-workers (3.4–3.7). The symmetry of armchair and zig-zag nanotubes is considered first. Such tubes can be represented by *symmorphic* groups, that is groups in which the rotations can be treated by simple point group representations. This differentiates them

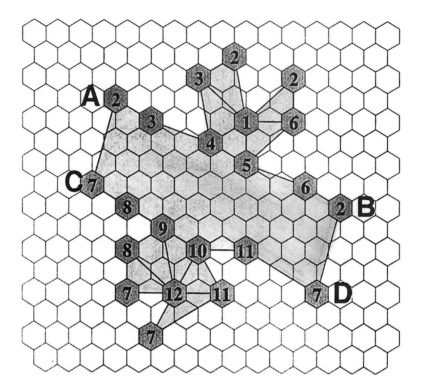

Fig. 3.12. Projection map illustrating two different ways of capping the chiral nanotube defined by the vector (7,5) (3.13).

from chiral tubes, in which the symmetry operations involve both translations and rotations.

In determining the symmetry classification of nanotubes, we assume that the tube length is much greater than the diameter, so that the caps can be neglected. Since all armchair and zig-zag nanotubes have a rotational symmetry axis and may additionally have either a mirror plane at right angles to this axis or an inversion centre, they fall into either the D_{nh} or the D_{nd} group. In deciding between these two groups, we follow Dresselhaus and colleagues in making the assumption that all tubes have an inversion centre. Now inversion is an element of D_{nh} only for even n, and is an element of D_{nd} only for odd n. It follows that the symmetry group for armchair or zig-zag tubes with n even is D_{nh} while the group for armchair or zig-zag tubes with n odd is D_{nd}.

For chiral tubes the symmetry groups are non-symmorphic. Thus the basic symmetry operation $R = (\psi,\tau)$ involves a rotation by an angle ψ followed by a translation τ. This operation corresponds to the vector $\mathbf{R} = p\mathbf{a}_1 + q\mathbf{a}_2$. Thus (p,q) denotes the coordinates reached when the symmetry operation (ψ,τ) acts on an atom at $(0,0)$. Dresselhaus *et al.* show that the values of p and q are given by

$$mp - nq = d_H,$$

with the conditions $q < m/d_H$ and $p < n/d_H$. It can also be shown that the parameters ψ and τ are given by

$$\psi = 2\pi \frac{\Omega}{N d_H}$$

and

$$\tau = \frac{T d_H}{N},$$

where the quantity Ω is defined as

$$\Omega = \{p(m + 2n) + q(n + 2m)\}/(d_H/d_R)$$

with

$$d_R = \begin{cases} d_H & \text{if } n - m \text{ is not a multiple of } 3d_H \\ 3d_H & \text{if } n - m \text{ is a multiple of } 3d_H. \end{cases}$$

We are now in a position to consider the symmetry group for chiral nanotubes. Unlike armchair and zig-zag nanotubes, chiral tubes contain no mirror planes, and therefore belong to C symmetry groups. Considering first tubes with $d_H = 1$, the order of the rotational axis is equal to 2π divided by the number of rotation operations required to reach a lattice vector, i.e. $2\pi/\psi = N/\Omega$, so that the symmetry group becomes $C_{N/\Omega}$. For tubes with $d_H \neq 1$, the symmetry group is expressed as a direct product, $C_{d_H} \otimes C'_{N/\Omega}$ (the direct product of two groups is a new group whose elements are the products of the elements of the first group with those of the second).

We can illustrate the meaning of some of these parameters with reference to Fig. 3.13, which relates to the chiral $(4,2)$ nanotube. In this case, the parameters p and q are 1 and 0 respectively, so that $\mathbf{R}$ is the vector joining $(0,0)$ to $(1,0)$. If we now imagine the two-dimensional sheet being rolled up to form the tube, the line joining $(0,0)$ to $(1,0)$, $(2,0)$, etc. becomes a helix running around the cylinder. Eventually this line intersects a lattice point at a distance T along the tube. The length of the line at the point of intersection is then $N/d_H \times R$, where N is one half of the number of atoms in the unit cell and d_H is the highest common divisor of n and m. For the $(4,2)$ nanotube, $N = 28$, $d_H = 2$, so the length of the line at the point of intersection is $14R$. This is represented in Fig. 3.13 by the intersection of the line joining $(0,0)$ to $(1,0)$, $(2,0)$ . . . to a line drawn through the

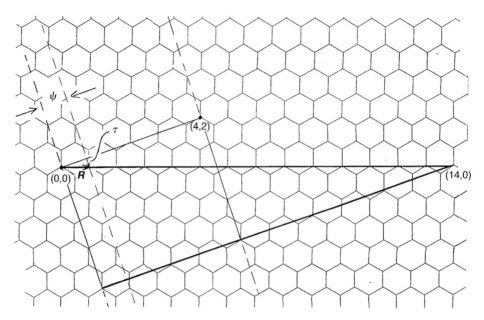

Fig. 3.13. Diagram illustrating symmetry operations for chiral nanotubes.

lower edge of the unit cell. The intersection occurs at the point denoted by (14,0). Now, for the (4,2) tube the quantity Ω is equal to 10, so that in this case the symmetry group is $C_2 \otimes C'_{28/10}$.

For a fuller discussion of the symmetry classification of nanotubes, the reader should consult Refs. (3.3)–(3.7) and other papers by Dresselhaus, Jishi and co-workers.

3.3.6 Elbow connections, tori and coils

Nanotubes containing abrupt elbow-like bends are occasionally observed experimentally (see Section 3.5.8). Connections of this type have been analysed by a number of workers (e.g. 3.15–3.20). It has been established that armchair tubes could be joined to zig-zag tubes by elbow connections involving a pentagonal ring on the outer side of the elbow and a heptagon on the inner side. As an example, Fig. 3.14 shows a connection between a (5,5) armchair tube and a (9,0) zig-zag tube (3.18). According to Dunlap (3.16), the optimal angle between tubes joined by a pentagon–heptagon connection should be 150°, but model-building exercises by Fonseca and colleagues (3.17) produced an angle of 144°. These workers point out that ten such connections in a common plane will result in a torus, as shown in Fig. 3.15(a). Nanotube tori, or 'hoops', have now been observed experimentally (see Section 3.6.3), although

(a) (b)

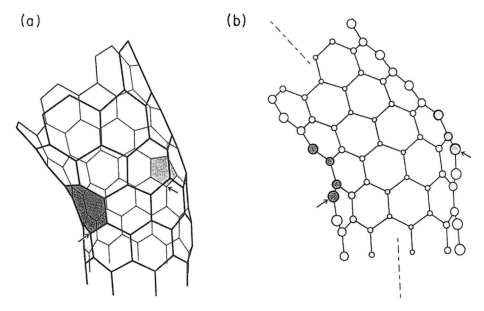

Fig. 3.14. Illustration of an 'elbow connection' between a (5,5) armchair and a (9,0) zig-zag nanotube (3.18). (a) Perspective drawing with pentagonal and heptagonal rings shaded, (b) structure projected on symmetry plane of elbow.

these have much larger diameters than the models shown here, and their structure has not yet been established in detail.

Figure 3.16(a) shows a planar representation of elbow connections between (9,0) and (5,5) nanotubes, again taken from the work of Fonseca *et al.* (3.17). If the position of the lower pentagon–heptagon pair is shifted by a single bond as shown in Fig. 3.16(b), the tube will no longer form a torus, and if this shift is repeated at regular intervals, the result will be a helix, or coil, as shown in Fig. 3.15(b). Such coiled tubes are often observed in catalytically produced nanotube samples, as discussed in Section 2.3.

3.3.7 Arrays of single-walled nanotubes

Single-walled nanotubes often occur in tightly packed bundles and this has prompted a number of authors to discuss the possible structures and properties of single-walled nanotube 'crystals'. Tersoff and Ruoff discussed hexagonally packed lattices of tubes with various diameters (3.21), and found a crossover between two distinct regimes as the tube size was increased. In arrays of tubes with diameters less than 1 nm, the individual tubes remained almost perfectly circular in cross-section, while tubes of 2.5 nm and above became flattened against one another as a result of van der Waals interactions,

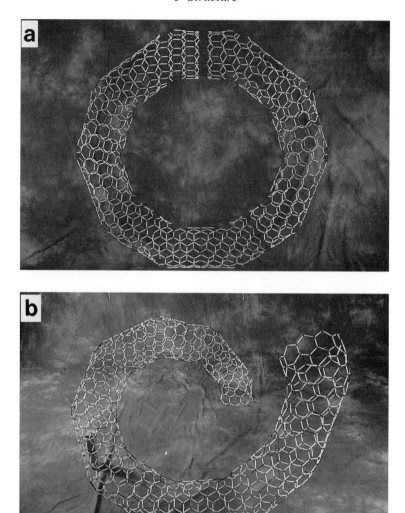

Fig. 3.15. (a) Model of a torus constructed from five (5,5) tube segments and five (9,0) tube segments connected by joints containing pentagonal and heptagonal rings. (b) One turn of a helically coiled tube constructed from the same segments as in (a), but with each segment rotated by a fixed angle (courtesy A. A. Lucas).

resulting in a honeycomb structure. Distortions of this type have been observed in experimental studies of adjacent multiwalled tubes (3.22).

Charlier and colleagues considered three alternative packings of (6,6) armchair tubes, which have a diameter of 0.814 nm (3.23). The crystalline structures they considered were: one tetragonal with space group $P4_2/mmc$ (D_{4h}^9) and two hexagonal with space groups $P6/mmm$ (D_{6h}^1) and $P6/mcc$ (D_{6h}^2)

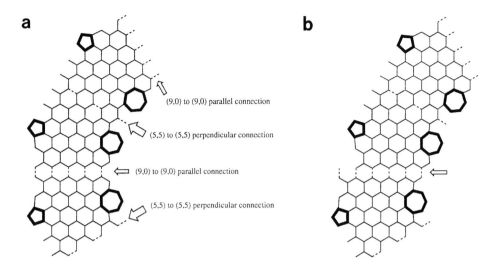

Fig. 3.16. (a) Planar representation of the connections which produce a torus. (b) Representation of a single rotational bond shift which leads to a rotation out of the plane, and a helical tube (3.17).

Fig. 3.17. Arrays of single-walled nanotubes, as considered by Charlier *et al.* (3.23). (a) Tetragonal arrangement, (b) hexagonal I and (c) hexagonal II arrangements.

respectively. These are illustrated in Fig. 3.17. The difference between the two hexagonal phases lies in the way in which atoms in neighbouring tubes line up with each other. In the hexagonal I phase, the atoms in neighbouring tubes are exactly aligned with each other, while in the hexagonal II phase only half the atoms have a corresponding atom in the neighbouring tube, with the other half lying next to the centre of a hexagon.

Charlier *et al.* determined the energies of the three lattices for a range of intertube distances. The hexagonal II phase was found to be the most stable structure, with an optimal intertube separation of 0.314 nm, followed by the hexagonal I lattice and the tetragonal phase, with optimal intertube distances of 0.335 nm and 0.336 nm respectively.

3.4 The physical stability of carbon nanotubes

The earliest work on the stability of carbon tubules as a function of diameter
was carried out by Gary Tibbetts in 1984 (3.24). Using a continuum model,
Tibbetts found that the strain energy of a thin graphitic tube varies with
1/(diameter). This implies that the strain energy per atom varies with 1/(diameter)^2. Following the discovery of fullerene-related nanotubes, a number of
groups have carried out more detailed theoretical studies of nanotube stability
(e.g. 3.25–3.27). John Mintmire and colleagues from the Naval Research Laboratory in Washington used empirical potentials to calculate the strain energies per carbon atom for all possible tubes with diameters less than 1.8 nm
(3.25). They also found that, to a good approximation, the strain energy per
atom varied with 1/(diameter)^2. For tubes with diameters larger than about
1.6 nm, the strain energy becomes very close to that in planar graphite. It is
interesting to note that this diameter is approximately the same as the smallest
observed experimentally in multiwalled nanotubes. Mintmire and colleagues
found that strain energy was independent of tube structure.

Other workers have taken a slightly different approach to determining the
stability of nanotubes (3.26, 3.27). This involves considering a nanotube as a
rolled-up graphene strip, and balancing the energy gained due to the elimination of edge atoms against the energy cost of bond-bending. In calculations of
this kind, Sawada and Hamada found that the critical diameter above which
tubes are more stable than strips is approximately 0.4–0.6 nm (3.26). Independent calculations by Lucas, Lambin and Smalley produced a similar result
(3.27).

3.5 Experimental studies of nanotube structure: multiwalled nanotubes

3.5.1 Techniques

Of the many techniques which have been used to study the structure of
nanotubes, high resolution transmission electron microscopy has undoubtedly
been the most useful. While X-ray diffraction can provide more accurate
crystallographic measurements, and scanning tunnelling microscopy can potentially give greater surface detail, only HREM can probe the internal structure of nanotubes, and thus reveal details of the stacking arrangement of
multiwalled tubes, the nature of defects, the structure of caps and so on. In the
discussion of nanotube structure which follows, most of the studies referred to
have employed high resolution electron microscopy. Electron diffraction has
also been applied to nanotubes, and is discussed in a separate section. Scann-

ing probe microscopy has proved rather more difficult to apply to nanotubes, partly because of the difficulty of attaching the tubes securely to suitable substrates. In the case of multiwalled nanotubes, high quality atomic-resolution images have proved very difficult to obtain, but some excellent images of single-walled tubes have now been achieved.

3.5.2 The layer structure: experimental observations

High resolution electron micrographs of multiwalled nanotubes generally show evenly spaced lattice fringes with an equal number of fringes on either side of the central core. Examples taken from Iijima's 1991 *Nature* paper (3.28) are shown in Fig. 3.18. However, in some cases it is found that the fringes on one or both sides of the cavity contain 'gaps', or anomalously large spacings, as shown in Fig. 3.19. The possible reasons for this are discussed in subsequent sections. Most X-ray diffraction measurements on nanotube samples give interlayer {0002} spacings of approximately 0.344 nm (e.g. 3.29, 3.30), which is very similar to the value found in turbostratic graphites, although figures ranging from 0.342 nm (3.11) to 0.375 nm (3.31) have been obtained. This large spread of values may result partly from the presence of gaps in the interlayer spacings, but may also reflect a variation in intershell distances with nanotube diameter. A more recent analysis of HREM images by Kiang and colleagues (3.32) has shown that the interlayer spacing can range from 0.34 nm to 0.39 nm depending on tube diameter, with smaller diameter tubes having the largest spacings. This was attributed to the high curvature of small tubes resulting in a greater repulsive intertube force.

A frequent observation in multiwalled tubes is the presence of one or more layers traversing the central core, as illustrated in Fig. 1.4. More complicated internal structures, sometimes involving the formation of closed compartments, are also quite commonly seen. An example is shown in Fig. 3.20. Such features can represent a barrier to the filling of nanotubes with foreign materials, as discussed in Chapter 5.

Scanning probe microscope studies can also sometimes give insights into the layer structures of multiwalled nanotubes. For example, work by Edman Tsang and colleagues from Oxford using atomic force microscopy (3.33) has shown that atomic-scale surface steps can be present on nanotube surfaces. Examples of their images are shown in Fig. 3.21. The presence of such steps suggests that defects are present in the multilayer structures, as discussed in the next section.

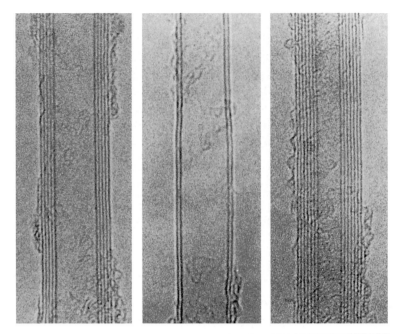

Fig. 3.18. Some of Iijima's first images of multiwalled nanotubes (3.28).

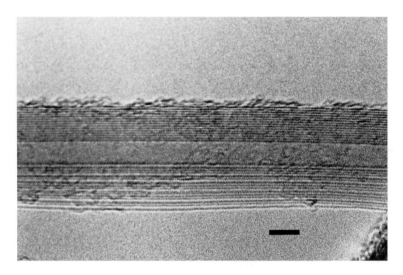

Fig. 3.19. Micrograph of multiwalled nanotube showing uneven spacings in the layer structure on either side of the central core. Scale bar 5 nm.

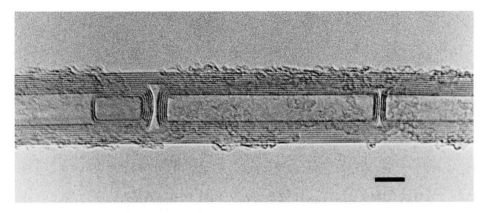

Fig. 3.20. Micrograph showing internal compartments in multiwalled nanotube. Scale bar 5 nm.

3.5.3 The layer structure: models

The most basic question concerning the layer structure of multiwalled tubes is whether they have a scroll-like, 'Swiss-roll', structure, or whether they instead consist of a 'Russian doll' arrangement of discrete tubes. These two possible arrangements are illustrated in Fig. 3.22. Alternatively, the structure might consist of a mixture of these two arrangements, as discussed by several authors (3.29, 3.34). Resolving this issue has proved rather difficult, and no consensus has yet been reached. The most direct way to determine the multilayer structure would be to image tubes end-on using high resolution electron microscopy, but this has not proved possible. The main problem is the extreme difficulty of aligning a tube parallel to the beam. Even if a tube could be correctly aligned, the length of the tube in the beam direction, and the likely presence of slight curvature, would make lattice imaging extremely difficult. An alternative possibility would be to prepare cross-sections of nanotubes, but this has not yet been achieved. Therefore, we have to infer the structure from more indirect measurements.

The fact that micrographs of multiwalled nanotubes almost invariably have an equal number of fringes on either side of the central cavity could be taken as providing support for the Russian doll model, although it certainly cannot be taken as unequivocal proof. Perhaps stronger evidence for a Russian doll structure is the presence in the tubes of internal caps (see Fig. 1.4) or closed compartments, as shown in Fig. 3.20. It is difficult to reconcile such features with a scroll structure. Further evidence that multiwalled nanotubes have a completely closed structure is provided by experiments on the opening and filling of nanotubes. As discussed in Chapter 5, a great deal of work has been

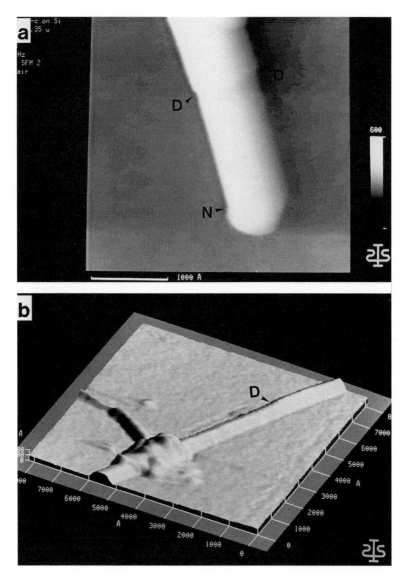

Fig. 3.21. AFM micrographs of multiwalled nanotubes (3.33). (a) Region near cap showing apparent edge dislocations (labelled D) and 'negative deviation' (N). (b) Body of tube, with another edge dislocation.

carried out on the reaction of nanotubes with gas phase or liquid phase oxidants, with the aim of opening the tubes. These studies invariably show that the tubes are preferentially attacked in the cap region, with the main body of the tubes being left intact. This seems to be inconsistent with a scroll model, which would have reactive surfaces all along the length of the tube, due to the terminating graphene layers.

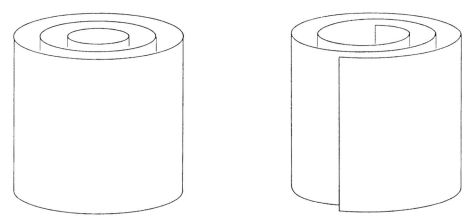

Fig. 3.22. Schematic illustration of 'Russian doll' and 'Swiss roll' models for multiwalled nanotubes.

Although the evidence in favour of a Russian doll structure seems quite
strong, arguments for alternative structures have been put forward by a
number of authors (e.g. 3.29, 3.34). Otto Zhou and colleagues, of Bell Labs
(3.29), have argued the case for a scroll-like structure, basing most of their
arguments on two sets of experiments. In the first they recorded X-ray diffraction patterns of nanotube samples subjected to high pressures in a diamond
anvil. At a pressure of 10.2 kbar they found that the {0002} spacing was
reduced from 0.344 nm to 0.340 nm, which is comparable to the *c*-axis com-
pressibility of graphite. In the second series of experiments they reacted the
nanotube samples with potassium and rubidium under conditions similar to
those which produce intercalation in C_{60}. In both cases they found weight
gains that implied the formation of intercalation compounds with the overall
composition MC_8. Zhou and colleagues use both sets of observations to
support their argument against the Russian doll model, although in the case of
the intercalation experiments this is open to doubt. Such experiments involve
the reaction of a nanotube sample with highly reactive metals, which might
cause major disruption of tubes with the Russian doll structure, enabling
intercalation to occur. The scroll-like model favoured by Zhou *et al.*, which
they describe as a 'papier mâché' structure, is illustrated in Fig. 3.23(a).

Severin Amelinckx and co-workers from the University of Antwerp have
also argued in favour of a structure containing scroll-like elements (3.34). Their
conclusions have been based chiefly on a detailed analysis of HREM images
recorded perpendicular to the tube axis. As already noted above, they point
out that the graphitic layers making up the nanotube walls often contain gaps,
as shown in Fig. 3.19, and conclude from this that the tubes must be at least

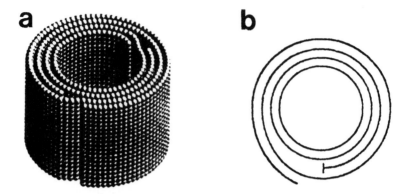

Fig. 3.23. (a) Defective scroll, or 'papier mâché', structure for multiwalled nanotubes, suggested by Zhou *et al.* (3.29). (b) Model by Amelinckx *et al.* for multiwalled tube containing both concentric shell and scroll structures (3.34).

partially scroll-like. Figure 3.23(b) illustrates, in cross-section, the kind of structure they envisage for multiwalled nanotubes, which contains both Russian doll and scroll-like elements. This structure contains edge-type dislocations, which Amelinckx *et al.* contend explain the presence of the gaps in HREM images recorded perpendicular to the tube axis. They argue that at the high temperatures used to produce nanotubes, the attractive van der Waals forces between adjacent layers would be less important than the repulsive forces, so that the gaps would not close. However, while this may be true at high temperatures, it seems unlikely that the gaps would remain open during the contraction which would occur during cooling, bearing in mind the known flexibility of the graphene layers. Moreover, dislocations similar to those described by Amelinckx *et al.* have been observed in nanoparticles (3.35) and these are not associated with gaps. An alternative explanation for the gaps is that they result from variations in the cross-sectional shape of nanotubes, as discussed in Section 3.5.6.

In summary, the balance of evidence seems to favour a Russian doll model for the structure of multiwalled nanotubes, but it is quite possible that this type of tube may co-exist with other structures such as those envisaged in the papers of Zhou and Amelinckx and their colleagues.

3.5.4 Electron diffraction

Iijima included electron diffraction patterns of individual nanotubes in his original *Nature* paper (3.28), and a number of groups have subsequently attempted to obtain structural information on nanotubes in this way. However, the patterns can be difficult to interpret, particularly with multiwalled tubes, as now discussed.

A flat graphite sheet oriented perpendicular to the electron beam would give a hexagonal diffraction pattern. If we now consider a beam impinging on a multiwalled nanotube in which all the tubes have an identical structure, as represented in Fig. 3.24(a), a similar pattern will result (3.10, 3.36). However, in this case we can tilt the beam in a direction perpendicular to the tube axis and the pattern will remain the same. Therefore, in reciprocal space the spots become circles situated in planes normal to the nanotube axis, as shown in Fig. 3.24(b). Since the stacking in such a tube would be turbostratic rather than of the ABAB . . . type, the spots will be streaked along a direction perpendicular to the tube axis. Further details of the reciprocal space construction for both chiral and non-chiral tubes are given in the papers of Zhang *et al.* (3.10, 3.36). Diffraction from nanotubes has also been discussed by Lambin, Lucas and colleagues (e.g. 3.37).

Experimental electron diffraction patterns of multiwalled nanotubes usually show spots from both non-chiral and chiral tubes. This could be taken as further evidence for a Russian doll structure, since a 'scroll' nanotube would have the same helicity throughout. A typical diffraction pattern, taken from the work of Zhang *et al.*, is shown in Fig. 3.25. This pattern was recorded from a multiwalled nanotube with 18 individual layers and an innermost diameter of 1.3 nm; the incident electron beam is normal to the tube axis. In this pattern the $\{000l\}$ spots which result from the parallel graphene layers perpendicular to the beam can be seen running horizontally on either side of the central spot. The other arrowed spots all correspond to reflections from achiral (i.e. zig-zag or armchair) tubes, and are of the $\{10\bar{1}0\}$ or $\{11\bar{2}0\}$ type. The other, weaker reflections are due to chiral tubes. Careful analysis of the spots from chiral nanotubes enables the chiral angle to be determined, but it is essential in such experiments that the tube is precisely aligned perpendicular to the electron beam.

3.5.5 Plan-view imaging by HREM

In addition to the strong $\{0002\}$ fringes present in all high resolution images of nanotubes, faint fringes with a finer spacing can sometimes be seen in the 'core' region of the tubes. An image of this kind, taken from the work of Zhang *et al.* (3.10), is shown in Fig. 3.26. Here, fringes with a spacing of 0.21 nm can be seen in parts of the central region. These fringes make a hexagonal pattern, which apparently suggests that the graphene sheets in successive tubes are parallel and that all the tubes are achiral. In most cases, however, the 'plan-view'

a

b

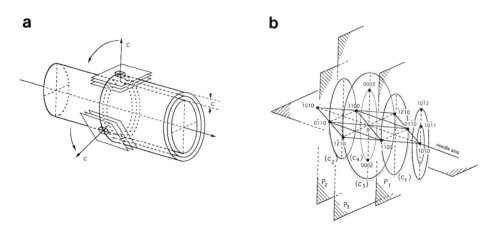

Fig. 3.24. (a) Schematic drawing of multiwalled nanotube illustrating graphite c-planes which are locally tangent to successive cylinders. (b) Reciprocal space construction for achiral tube. The loci of the reciprocal lattice points of graphite are circles (C_i) situated in planes (P_i) perpendicular to the tube axis (3.36).

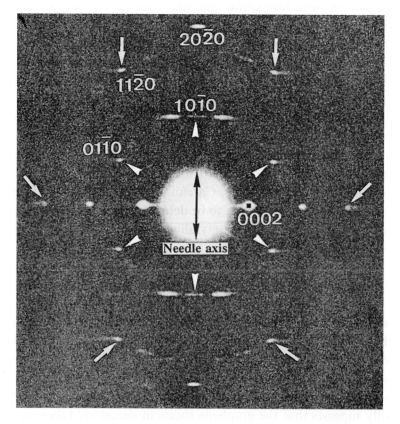

Fig. 3.25. Electron diffraction pattern of a multiwalled nanotube (3.36).

fringes in images of multiwalled nanotubes are difficult to interpret, which is not surprising since the individual tubes probably have a variety of structures and helicities.

3.5.6 The cross-sectional shape of multiwalled nanotubes

Direct observations of the cross-sectional shape of nanotubes in the electron microscope have proved very difficult to achieve, for the reasons mentioned above. Therefore, inferences about the cross-sectional shape of multiwalled tubes have to be made from images recorded perpendicular to the tube axis. It is also important to consider the influence of cap structure on the likely cross-sectional shapes. For example, Ebbesen has argued that a cap with five-fold symmetry will impose a faceted shape on the tube (3.38). However, this effect is likely to be slight, and will diminish as one moves further away from the cap region. A stronger effect on nanotube shape is probable with 'asymmetric cone' caps, of the kind discussed in Section 3.5.7 (Fig. 3.29). Caps of this type will result in an 'egg shaped' cross-section, although once again this effect may diminish in regions well removed from the caps.

It was noted above that high resolution electron micrographs of multiwalled nanotubes often show unevenly spaced lattice fringes on one or both sides of the core, as in Fig. 3.19. Amelinckx and colleagues suggested that this may be due to scroll-like elements in the multilayer structure, but an alternative explanation has been proposed by Mingqi Liu and John Cowley of Arizona State University (3.39). These workers carried out a detailed analysis of high resolution images of multiwalled nanotubes, and concluded that in many cases the tubes had polygonal cross-sections made up of flat regions joined by regions of high curvature. In images of the regions where the two planar sheets join, Liu and Cowley argued that fringes with spacings greater than 0.34 nm would be observed. This situation is illustrated schematically in Fig. 3.27. Note that their model assumes a relatively 'perfect' multilayer structure, and the observed gaps are a consequence of the joining together of idealised flat regions with the 0.34 nm spacing. The maximum interlayer spacing predicted by this model is 0.41 nm. In reality spacings considerably larger than this are observed (in Fig. 3.19 spacings greater than 0.6 nm are present). In such cases the multilayer structure is probably rather more imperfect than envisaged by Liu and Cowley.

Another factor which must be taken into account when considering the cross-sectional shapes of nanotubes is the presence of distorting forces resulting from contacts between adjacent tubes. This effect was discussed by Rodney Ruoff and colleagues (3.22), whose work was mentioned above in connection with the structure of nanotube arrays. These workers described high resolution

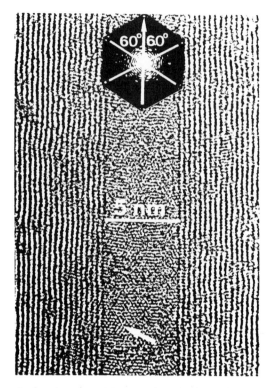

Fig. 3.26. High resolution image with 0.21 nm spacing fringes visible in the 'core' region. An optical transform from the core region is inset (3.10).

electron microscopy observations of nanotubes in contact along one edge. In micrographs of adjacent nanotubes, {0002} fringes were found to be more intense along the inner region, where the two tubes make contact, than at the outer edges, indicating a flattening of the tubes along the contact region. Ruoff *et al.* carried out calculations for the interaction of a pair of double-layer tubes using a Lennard-Jones model for the van der Waals interaction. These showed considerable flattening in the contact area, in agreement with the experimental observations. Ruoff and colleagues also found that the {0002} interlayer spacings on the sides of the tubes adjacent to the contact region were reduced by about 0.008 nm compared with those on the outer sides.

3.5.7 HREM studies of cap structure

Theoretical work on the capping of nanotubes was discussed in Section 3.3.4, where it was noted that tubes with diameters larger than about 1 nm can be capped in a large number of different ways. Experimental studies show that

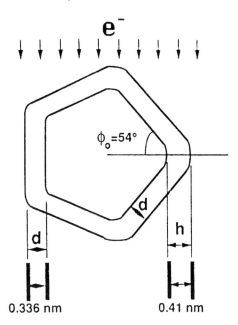

Fig. 3.27. Schematic drawing of polygonalised nanotube, as envisaged by Liu and Cowley (3.39).

multilayer nanotube caps do indeed have a wide range of different structures. In the great majority of cases the cap structures are unsymmetrical, but caps with higher symmetry are sometimes seen. Two beautiful micrographs, by Iijima (3.40), of symmetrical tube caps are shown in Fig. 3.28, together with diagrams indicating the approximate positions of the pentagonal rings in each case. It is notable that the degree of faceting increases as one moves from the inner graphene layers to the outer layers. This is in general agreement with predictions about the shapes of higher fullerenes, namely that they become less spherical as the size increases. Iijima estimates that the largest cap of the left-hand tube corresponds to one half of the icosahedral fullerene C_{6000}.

A commonly observed type of nanotube cap is the 'asymmetric cone' structure, illustrated in Fig. 3.29. This type of structure is believed to result from the presence of a single pentagon at the position indicated by the arrow, with five further pentagons at the apex of the cone. Theory predicts that the cone angle produced by five pentagons should be 19.2° (see Table 3.1). In practice, the angles observed in asymmetric cone caps can differ quite significantly from this value. In the case of the cap shown in Fig. 3.29, the angle is approximately 26°. As noted above, this type of cap imposes a non-circular cross-sectional shape on the nanotube. Rather less common are caps displaying a 'bill-like' morphology such as that shown in Fig. 3.30 (3.41). This structure results from the

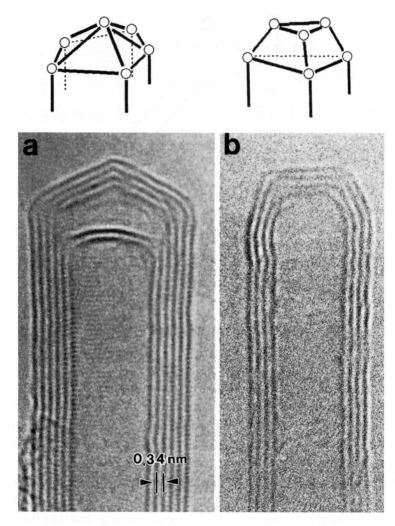

Fig. 3.28. Micrographs showing symmetrical nanotube caps, with drawings indicating location of pentagons (3.40).

presence of a single pentagon at point A and a heptagon at point B (see Fig. 1.6 for another structure of this type).

Although virtually all multiwalled nanotubes in samples produced by arc-evaporation are closed, examples are sometimes observed which are complete-ly open, with no obvious cap structure. An example is shown in Fig. 3.31(a). Careful analysis of such structures by Iijima and colleagues (3.40, 3.42) has demonstrated that in such cases the tubes are terminated with semitoroidal structures containing six pentagon–heptagon pairs, as illustrated in Fig. 3.31(b). The walls consist of successive folded graphene sheets, with no dangl-

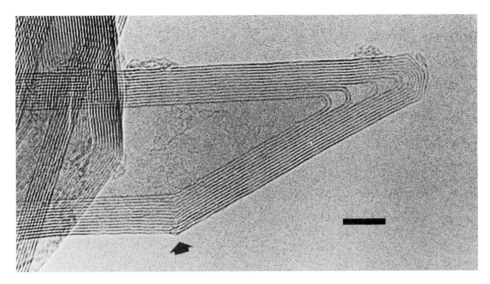

Fig. 3.29. Nanotube cap with asymmetric cone structure. Scale bar 5 nm.

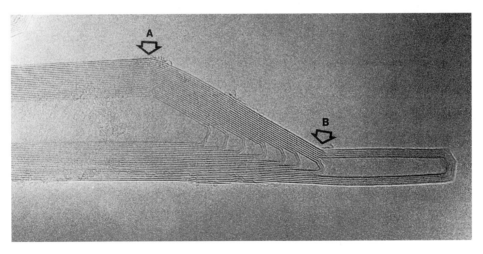

Fig. 3.30. Nanotube cap with bill-like structure (3.41).

ing edges. Occasionally, more complex structures are seen in which an inner tube extends beyond a semitoroidal tube termination (3.40).

3.5.8 Elbow connections and branching structures

As noted above, nanotubes containing abrupt elbow-like bends are sometimes observed in samples prepared using the conventional arc-evaporation method. An example is shown in Fig. 3.32. In most cases the tubes on either side of the

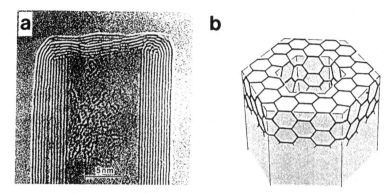

Fig. 3.31. (a) Micrograph of semitoroidal tube termination. (b) Schematic drawing of structure (3.40).

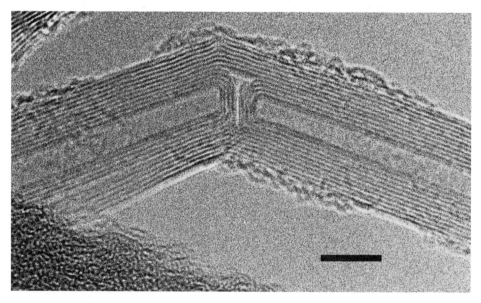

Fig. 3.32. Elbow connection joining two multiwalled nanotubes. Scale bar 5 nm.

elbow are of different diameters, and the joints are frequently with 'internal caps'. It has been suggested that this type of structure results from the presence of a pentagonal ring on the outer side of the elbow and a heptagon on the inner side, as discussed in Section 3.3.6. However, in experimental images, there are often discontinuities in the layer structure on the inner side of the elbow joint, as can be seen in Fig. 3.32. Therefore, the structure of the connections may be less perfect than envisaged by theoreticians. Measurement of the exact angle on electron micrographs is difficult because it is almost impossible to know

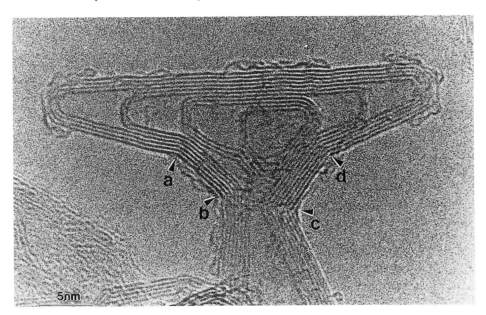

Fig. 3.33. Example of branching nanotube structure, produced by arc-evaporation method with modified electrodes (3.43).

whether the tube axis is precisely perpendicular to the electron beam direction. However, angles differing quite considerably from 150° have been measured, even where the joint is believed to be approximately perpendicular to the beam.

More complex, branching nanotube structures, which also apparently contained negative curvature, were observed in 1995 by Dan Zhou and Supapan Seraphin of the University of Arizona (3.43). These structures were produced under the usual arc-evaporation conditions, but with a graphite anode which had been drilled out to leave a hollow core approximately 0.32 cm in diameter. Three types of branched structure were described, with 'L', 'Y' and 'T' configurations. An example of the T type is shown in Fig. 3.33 with the negatively curved points labelled a, b, c and d. As with the elbow connections, the angles made by these junctions varied quite considerably. The reason for the formation of such structures is not clear at present.

Elbow connections and branching structures of the kind described here could be thought of as constituting the first steps towards building 'molecular scaffolding' from nanotubes (see concluding chapter). Elbow junctions might also have interesting electronic properties, as discussed in Chapter 4 (Section 4.2.4).

3.6 Experimental studies of nanotube structure: single-walled nanotubes

3.6.1 High resolution electron microscopy and electron diffraction

Samples of single-walled nanotubes tend to be much more homogeneous than samples of multiwalled nanotubes, with a much narrower range of diameters, and contain fewer obvious defects. High resolution electron micrographs of single-walled nanotubes generally show 'featureless' narrow tubes, and are therefore relatively uninformative compared with images of multiwalled nanotubes. Some of the general features of single-walled nanotubes were summarised in Section 2.5. The tubes often form bundles, and images of these bundles viewed end-on show close-packed arrays of tubes, as shown in Fig. 2.19. Both the tube bundles and individual tubes are frequently curled and looped. However, regular helical structures such as those observed in catalytically produced multiwalled tubes are not observed in single-walled nanotubes. The caps of single walled nanotubes, like those of multiwalled tubes, can have various shapes although most of them appear to be simple domes. Asymmetric cone caps are quite common; an example can be seen at the bottom left of Fig. 2.18. Elbow connections appear to be rare in single-walled tubes, but Iijima has observed some striking structures involving changes in the growth direction, as shown in Fig. 3.34 (3.44). The first of these, Fig. 3.34(a), can be understood in terms of pentagonal and heptagonal defects in the positions indicated, while the second 'candy-cane' structure, Fig. 3.34(b), may result from pentagon–heptagon pairs such as those which are present in the semitoroidal caps discussed above.

Recording electron diffraction patterns from single-walled nanotubes is exceptionally difficult owing to their very small diameters and beam-sensitivity. Iijima has reported obtaining patterns from individual tubes (3.44, 3.45) and in each case found the tubes to be helical. Diffraction patterns can more readily be obtained from bundles of single-walled nanotubes, and several groups have carried out such studies. John Cowley of Arizona State University, in collaboration with the Smalley group used a scanning transmission electron microscope (STEM) to obtain electron diffraction patterns of nanotube rope samples (3.46). Patterns were recorded from regions about 0.7 nm in diameter and indicated that, in most cases, the individual tubes had zero helicity, consistent with the (10,10) armchair structure suggested by Thess and colleagues in their original 'ropes' paper (3.47). However, subsequent STM studies (see next section) suggested that most of the tubes had chiral structures.

Iijima and colleagues have also described electron diffraction studies of bundles of single-walled nanotubes (3.48). In general they did not find any

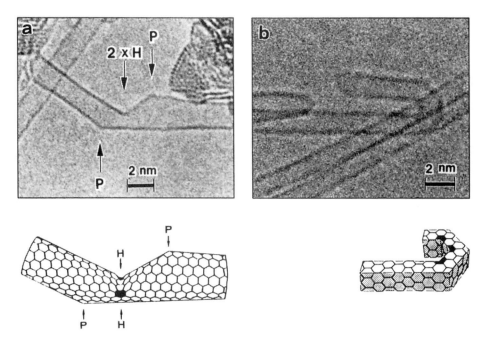

Fig. 3.34. Single-walled nanotubes displaying changes in growth direction, with drawings showing positions of heptagons and pentagons, from the work of Iijima (3.44).

preferred helicities in these groups of tubes, although in one case a bundle was found which apparently consisted almost entirely of armchair tubes.

3.6.2 Scanning probe microscopy

As already noted, achieving atomic resolution STM and AFM images of carbon nanotubes is fraught with difficulty. Quite apart from the practical problems, pitfalls can arise in image interpretation. This is particularly the case when highly oriented pyrolytic graphite (HOPG) is employed as a substrate. Scanning tunnelling microscopy images of HOPG are notoriously subject to artefacts, as illustrated by the controversy over early images of DNA on HOPG (3.49). The risk of misinterpretation is particularly acute with nanotubes, since the atomic structure here is identical with that of HOPG. For these reasons, there are advantages in using substrates other than HOPG for studies of carbon nanotubes.

In early 1998, two groups described atomic resolution STM studies of nanotube 'rope' samples prepared by laser vaporisation (3.50, 3.51). Both groups used single-crystal gold as the substrate. In addition to imaging the samples, electronic measurements on individual tubes were made using scanning tunnelling spectroscopy; these are discussed in Chapter 4. In contrast to previous studies, the atomic resolution STM images showed that most of the tubes had chiral structures. Examples of these STM images, from the work of Charles Lieber and colleagues of Harvard (3.51) are shown in Fig. 3.35. The tube shown in (a) has a chiral angle of approximately 8°, and a diameter of ~ 1 nm, which Lieber *et al.* interpreted as suggesting either a (11,2) or a (12,2) structure. The lower tube in (b) was found to have a chiral angle of about 11°, and a diameter of ~ 1.08 nm, suggesting a (12,3) structure. The earlier studies, using electron diffraction and other techniques, had suggested that the rope samples contained a high proportion of (10,10) armchair tubes, so the STM results showing chiral structures came as something of a surprise. One possibility, suggested by the Harvard group, is that isolated tubes, which were most likely to be imaged in the STM experiments may have a different structure from those in the ropes themselves. Overall, the structure of nanotubes in rope samples remains uncertain at the present time.

3.6.3 Nanotube hoops and diameter doubling

In the course of examining nanotube rope samples using transmission electron microscopy and scanning force microscopy, Smalley's group frequently observed circular formations such as that shown in Fig. 3.36 (3.52). Initially sceptical that these structures could be made up of seamless nanotubes, they christened them 'crop circles'. However, detailed examination suggested that the circles were indeed seamless nanotube tori. The SFM image in Fig. 3.36 shows a hoop apparently consisting of an individual nanotube. The formation mechanism of such structures is unclear, but presumably must involve the collision of a growing tube tip with its own tail. As Smalley and colleagues point out, this recalls Kekulé's vision of a chain of carbon atoms forming a benzene ring. The curvature of the growing tube may result from the presence of pentagon–heptagon pairs, as discussed in Section 3.3.6 above. With their unique structures, nanotube hoops may have interesting electronic properties, and these have been discussed by several authors (e.g. 3.53, 3.54).

Another fascinating observation made by Smalley and colleagues concerned the effect of heat treatment on nanotube rope samples (3.55). In samples annealed at 1400–1500 °C in vacuum or in flowing argon or hydrogen, a high proportion of the tubes were found to have diameters twice (2.7 nm) or occasionally three times (4.1 nm) the diameters of the original tubes. This was attributed to the coalescence of adjacent tubes, and provides clear evidence that large diameter tubes are energetically favoured over narrower ones. Smalley *et al.* put forward a mechanism for tube coalescence in which defects in

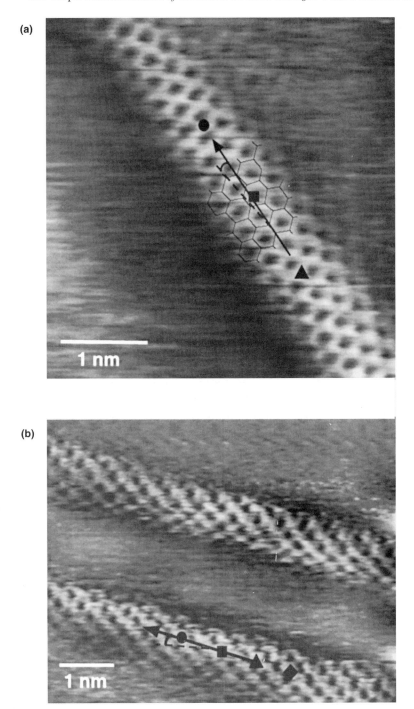

Fig. 3.35. Scanning tunnelling microscope images of single-walled nanotubes in 'rope' sample (3.51). (a) Individual tube exposed at surface of rope, (b) isolated tubes.

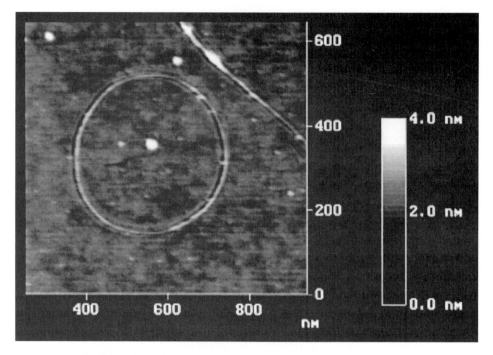

Fig. 3.36. Atomic force microscope image of 'nanohoop' (3.52).

the tube sides are initially produced by reaction with gas phase species, and the tubes then knit together with a 'zipping' action. They argued that this will only be possible if the adjacent tubes have identical helicities.

It is well established that small fullerene molecules can coalesce to form larger ones. The observation by Wang and Buseck of elongated fullerenes in crystals of fullerite was mentioned in Chapter 1. Such structures probably arose from the coalescence of adjacent C_{60} or C_{70} molecules in the crystal. Work by the Rice group in 1991 indicated that metallofullerenes containing two to four metal atoms could form by the coalescence of smaller metallofullerenes containing just one atom (3.56). In subsequent work the apparent coalescence of empty fullerenes was observed (3.57). More recently, Japanese workers have shown that heat treatment of crystalline C_{60} at very high temperatures (up to 2400 °C) resulted in the formation of fullerene-like nanoparticles with diameters in the size range 5–15.0 nm (3.58).

3.7 Structure of carbon nanoparticles

As discussed in the previous chapter, carbon nanotubes produced by the arc-evaporation method are invariably accompanied by graphitic nanopar-

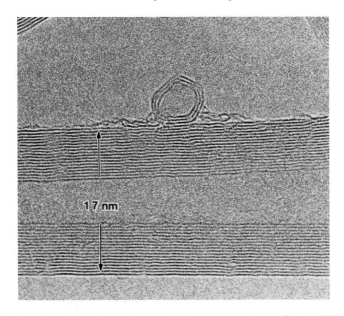

Fig. 3.37. Image of small carbon nanoparticle on nanotube surface (3.59). The particle has three concentric layers.

ticles. These have not been subjected to the level of detailed analysis that has been applied to nanotubes, but are probably best thought of as having structures related to large, rather imperfect multilayered fullerenes. As in the case of nanotubes, it is not entirely clear whether the structures are concentric or whether they have a spiralling, nautilus-shell structure. High resolution images of nanoparticles rarely provide unequivocal evidence one way or the other. Evidence that nanoparticles have relatively imperfect structures is provided by the fact that they can be preferentially removed from samples of cathodic soot by intercalation and oxidation (see Section 2.7). This seems to indicate that most nanoparticles do not have closed structures. Also, high resolution images of nanoparticles often show discontinuities, which have the appearance of line dislocations (3.35). On the other hand, images of very small nanoparticles sometimes clearly show concentric Russian doll structures. An example, taken from the work of Ando and Iijima is shown in Fig. 3.37 (3.59). Thus, it appears that carbon nanoparticles can be either closed or nautilus-like.

The nanoparticles exhibit a wide variety of different shapes, although the outer shape seems to depend on the size of the inner cavity. Thus, nanoparticles with large central cores tend to be more faceted than those with small inner shells. The faceted particles sometimes have pentagonal or hexagonal profiles, but are more frequently irregularly shaped with more than six sides. Figure 3.38 shows a typical example.

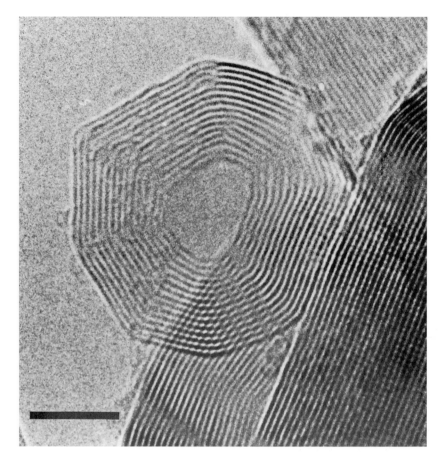

Fig. 3.38. Image of a typical faceted nanoparticle. Scale bar 5 nm.

In contrast with the situation for carbon nanotubes, there have been few, if any, theoretical discussions devoted specifically to the structure of carbon nanoparticles. However, as already noted, it seems clear that they are structurally related to large fullerenes. The shapes of giant fullerenes have been considered by Humberto and Mauricio Terrones (3.60), and some of their proposed structures are shown in Fig. 3.39. As mentioned above, nanoparticles with well-defined shapes are sometimes seen, and these probably have structures similar to those shown in these simulations.

3.8   Nanocones

There have been a number of studies of carbon materials, produced by a variety of methods, in which cone-shaped graphitic structures have been

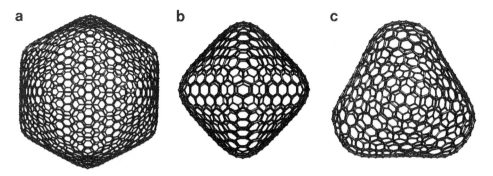

Fig. 3.39. Giant fullerene structures with various symmetries (3.60). (a) C_{1500} (I_h symmetry), (b) C_{600} (D_{2h} symmetry), (c) C_{660} (tetrahedral symmetry).

observed (e.g. 3.61, 3.62). The most striking experiments in this area were carried out by Thomas Ebbesen and a group of colleagues from France, the US and Norway in 1997 (3.62). These workers prepared carbon samples by the pyrolysis of hydrocarbons using a carbon arc plasma generator. They estimated that the temperature of the plasma during pyrolysis was at least 2000 °C. Electron microscopy showed that the solid carbon produced in this way contained a large number of conical structures, as well as disks, open nanotubes and disordered material. Examples are shown in Fig. 3.40. The cone angles appeared to correspond to the five possible angles of cones containing one to five pentagonal rings (see Section 3.3.4). Ebbesen and colleagues discuss these results in terms of the formation of fullerene-related structures from monocyclic carbon precursors.

3.9 Discussion

In the early days of nanotube research, the beautiful images published by Iijima and others gave the impression of structures with a high degree of perfection. Subsequent work has tended to show that real nanotubes are less perfect in structure than was initially thought. For example, the cross-sectional shape of multiwalled tubes frequently appears to be non-circular, and irregularly spaced gaps are often present among the layers. Elbow connections are commonly observed, and under certain conditions more complicated branching structures can be formed. Other defects may also be present. For example, Thomas Ebbesen has suggested that pentagon–heptagon pairs may occur frequently along the length of multiwalled tubes (e.g. 3.63). There is evidence that some of these imperfections can be removed by annealing at high temperatures, but it remains to be seen whether modifications to the production technique can help

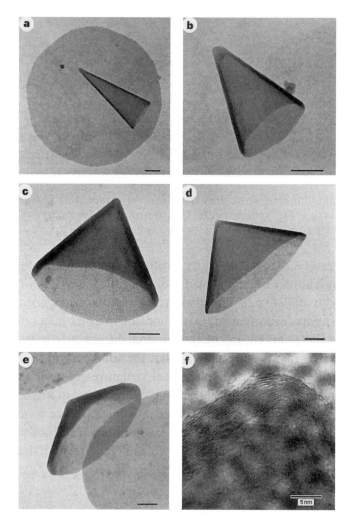

Fig. 3.40. Transmission electron micrographs of carbon nanocones (3.62). The cone angles, α, are as follows: (a) 19.2°, (b) 38.9°, (c) 60°, (d) 84.6°, (e) 112.9°; (f) shows a high resolution image of a cone tip.

to form multiwalled nanotubes with a higher degree of perfection than is currently observed.

Single-walled nanotubes generally appear to have more perfect structures than their multiwalled counterparts, and for this reason have often been preferred for experiments on the electronic properties such as those described in Chapter 4. However, structural studies of single-walled tubes are still at quite an early stage, and these structures could prove to be less perfect than currently believed. Certainly, micrographs such as those shown in Fig. 3.34 are evidence that single-walled tubes can contain imperfections.

The structures of carbon nanoparticles have not yet been explored in detail, and further work would be useful here, both from a fundamental standpoint and in view of the potential applications of these structures. Studies of other structures such as hoops and cones are still in their infancy. In general, there is considerable scope for further experimental work on the structures of carbon nanotubes and related materials. Quite possibly, such studies might lead to the discovery of yet more fascinating carbon architectures.

References

(3.1) K. R. Leonard, A. K. Kleinschmidt, N. Agabian-Keshishian, L. Shapiro and J. V. Maizel, 'Structural studies on the capsid of *Caulobacter crescentus* bacteriophage φCbK', *J. Mol. Biol.*, **71**, 201 (1972).

(3.2) R. O. Erickson, 'Tubular packing of spheres in biological fine structure', *Science*, **181**, 705 (1973).

(3.3) R. Saito, M. Fujita, G. Dresselhaus and M. S. Dresselhaus, 'Electronic structure of chiral graphene tubules', *Appl. Phys. Lett.*, **60**, 2204 (1992).

(3.4) M. S. Dresselhaus, G. Dresselhaus and R. Saito, 'Physics of carbon nanotubes', *Carbon*, **33**, 883 (1995).

(3.5) M. S. Dresselhaus, G. Dresselhaus and P. C. Eklund, *Science of fullerenes and carbon nanotubes*, Academic Press, San Diego, 1996.

(3.6) R. A. Jishi, M. S. Dresselhaus and G. Dresselhaus, 'Symmetry properties of chiral carbon nanotubes', *Phys. Rev. B*, **47**, 16,671 (1993).

(3.7) R. A. Jishi, D. Inomata, K. Nakao, M. S. Dresselhaus and G. Dresselhaus, 'Electronic and lattice properties of carbon nanotubes', *J. Phys. Soc. Japan*, **63**, 2252 (1994).

(3.8) C. T. White, D. H. Robertson and J. W. Mintmire, 'Helical and rotational symmetries of nanoscale graphitic tubules', *Phys. Rev. B*, **47**, 5485 (1993).

(3.9) J. W. Mintmire and C. T. White, in *Carbon nanotubes: preparation and properties*, ed. T. W. Ebbesen, CRC Press, Boca Raton, 1997, p. 191.

(3.10) X. F. Zhang, X. B. Zhang, G. Van Tendeloo, S. Amelinckx, M. Op de Beeck and J. Van Landuyt, 'Carbon nano-tubes; their formation process and observation by electron microscopy', *J. Cryst. Growth*, **130**, 368 (1993).

(3.11) D. Reznik, C. H. Olk, D. A. Neumann and J. R. D. Copley, 'X-ray powder diffraction from carbon nanotubes and nanoparticles', *Phys. Rev. B*, **52**, 116 (1995).

(3.12) J.-C. Charlier and J.-P. Michenaud, 'Energetics of multilayered carbon tubules', *Phys. Rev. Lett.*, **70**, 1858 (1993).

(3.13) M. Fujita, R. Saito, G. Dresselhaus and M. S. Dresselhaus, 'Formation of general fullerenes by their projection on a honeycomb lattice', *Phys. Rev. B*, **45**, 13,834 (1992).

(3.14) M. S. Dresselhaus, G. Dresselhaus and P. C. Eklund, 'Fullerenes', *J. Mater. Res.*, **8**, 2054 (1993).

(3.15) L. A. Chernozatonskii, 'Carbon nanotube elbow connections and tori', *Phys. Lett. A*, **170**, 37 (1992).

(3.16) B. I. Dunlap, 'Relating carbon tubules', *Phys. Rev. B*, **49**, 5643 (1994).

(3.17) A. Fonseca, K. Hernadi, J. B. Nagy, P. Lambin and A. A. Lucas, 'Model structure of perfectly graphitizable coiled carbon nanotubes', *Carbon*, **33**, 1759 (1995).

(3.18) P. Lambin, A. Fonseca, J. P. Vigneron, J. B. Nagy and A. A. Lucas, 'Structural and electronic properties of bent carbon nanotubes', *Chem. Phys. Lett.*, **245**, 85 (1995).

(3.19) P. Lambin, J. P. Vigneron, A. Fonseca, J. B. Nagy and A. A. Lucas, 'Atomic structure and electronic properties of bent carbon nanotubes', *Synth. Met.*, **77**, 249 (1996).

(3.20) L. Chico, V. H. Crespi, L. X. Benedict, S. G. Louie and M. L. Cohen, 'Pure carbon nanoscale devices: nanotube heterojunctions', *Phys. Rev. Lett.*, **76**, 971 (1996).

(3.21) J. Tersoff and R. S. Ruoff, 'Structural properties of a carbon nanotube crystal', *Phys. Rev. Lett.*, **73**, 676 (1994).

(3.22) R. S. Ruoff, J. Tersoff, D. C. Lorents, S. Subramoney and B. Chan, 'Radial deformation of carbon nanotubes by van der Waals forces', *Nature*, **364**, 514 (1993).

(3.23) J.-C. Charlier, X. Gonze and J.-P. Michenaud, 'First-principles study of carbon nanotube solid-state packings', *Europhys. Lett.*, **29**, 43 (1995).

(3.24) G. G. Tibbetts, 'Why are carbon filaments tubular?', *J. Cryst. Growth*, **66**, 632 (1984).

(3.25) D. H. Robertson, D. W. Brenner and J. W. Mintmire, 'Energetics of nanoscale graphitic tubules', *Phys. Rev. B*, **45**, 12,592 (1992).

(3.26) S. Sawada and N. Hamada, 'Energetics of carbon nanotubes', *Solid State Comm.*, **83**, 917 (1992).

(3.27) A. A. Lucas, P. Lambin and R. E. Smalley, 'On the energetics of tubular fullerenes', *J. Phys. Chem. Solids*, **54**, 587 (1993).

(3.28) S. Iijima, 'Helical microtubules of graphitic carbon', *Nature*, **354**, 56 (1991).

(3.29) O. Zhou, R. M. Fleming, D. W. Murphy, C. H. Chen, R. C. Haddon, A. P. Ramirez and S. H. Glarum, 'Defects in carbon nanostructures', *Science*, **263**, 1744 (1994).

(3.30) Y. Saito, T. Yoshikawa, S. Bandow, M. Tomita and T. Hayashi, 'Interlayer spacings in carbon nanotubes', *Phys. Rev. B*, **48**, 1907 (1993).

(3.31) M. Bretz, B. G. Demczyk and L. Zhang, 'Structural imaging of a thick-walled carbon microtubule', *J. Cryst. Growth*, **141**, 304 (1994).

(3.32) C.-H. Kiang, M. Endo, P. M. Ajayan, G. Dresselhaus and M. S. Dresselhaus, 'Size effects in carbon nanotubes', *Phys. Rev. Lett.*, **81**, 1869 (1998).

(3.33) S. C. Tsang, P. de Oliveira, J. J. Davis, M. L. H. Green and H. A. O. Hill, 'The structure of the carbon nanotube and its surface topography probed by transmission electron microscopy and atomic force microscopy', *Chem. Phys. Lett.*, **249**, 413 (1996).

(3.34) S. Amelinckx, D. Bernaerts, X. B. Zhang, G. Van Tendeloo and J. Van Landuyt, 'A structure model and growth mechanism for multishell carbon nanotubes', *Science*, **267**, 1334 (1995).

(3.35) P. J. F. Harris, M. L. H. Green and S. C. Tsang, 'High-resolution electron microscopy of tubule-containing graphitic carbon', *J. Chem. Soc., Faraday Trans.*, **89**, 1189 (1993).

(3.36) X. B. Zhang, X. F. Zhang, S. Amelinckx, G. Van Tendeloo and J. Van Landuyt, 'The reciprocal space of carbon tubes: a detailed interpretation of the electron diffraction effects', *Ultramicroscopy*, **54**, 237 (1994).

(3.37) P. Lambin and A. A. Lucas, 'Quantitative theory of diffraction by carbon nanotubes', *Phys. Rev. B*, **56**, 3571 (1997).

(3.38) T. W. Ebbesen, 'Carbon nanotubes', *Ann. Rev. Mater. Sci.*, **24**, 235 (1994).

(3.39) M. Liu and J. M. Cowley, 'Structures of carbon nanotubes studied by HRTEM and nanodiffraction', *Ultramicroscopy*, **53**, 333 (1994).

(3.40) S. Iijima, 'Growth of carbon nanotubes', *Mater. Sci. Eng. B*, **19**, 172 (1993).

(3.41) S. Iijima, T. Ichihashi and Y. Ando, 'Pentagons, heptagons and negative curvature in graphite microtubule growth', *Nature*, **356**, 776 (1992).

(3.42) S. Iijima, P. M. Ajayan and T. Ichihashi, 'Growth model for carbon nanotubes', *Phys. Rev. Lett.*, **69**, 3100 (1992).

(3.43) D. Zhou and S. Seraphin, 'Complex branching phenomena in the growth of carbon nanotubes', *Chem. Phys. Lett.*, **238**, 286 (1995).

(3.44) S. Iijima, 'TEM characterisation of graphitic structures', *Proc. 13th Int. Congr. Electron Microscopy*, Paris, 1994, p. 295.

(3.45) S. Iijima and T. Ichihashi, 'Single-shell carbon nanotubes of 1-nm diameter', *Nature*, **363**, 603 (1993).

(3.46) J. M. Cowley, P. Nikolaev, A. Thess and R. E. Smalley, 'Electron nano-diffraction study of carbon single-walled nanotube ropes', *Chem. Phys. Lett.*, **265**, 379 (1997).

(3.47) A. Thess, R. Lee, P. Nikolaev, H. Dai, P. Petit, J. Robert, C. Xu, Y. H. Lee, S. G. Kim, A. G. Rinzler, D. T. Colbert, G. E. Scuseria, D. Tománek, J. E. Fischer and R. E. Smalley, 'Crystalline ropes of metallic carbon nanotubes', *Science*, **273**, 483 (1996).

(3.48) L.-C. Qin, S. Iijima, H. Kataura, Y. Maniwa, S. Suzuki and Y. Achiba, 'Helicity and packing of single-walled carbon nanotubes studied by electron nanodiffraction', *Chem. Phys. Lett.*, **268**, 101 (1997).

(3.49) C. R. Clemmer and T. P. Beebe, 'Graphite – a mimic for DNA and other biomolecules in scanning tunnelling microscopy studies', *Science*, **251**, 640 (1991).

(3.50) J. W. G. Wildöer, L. C. Venema, A. G. Rinzler, R. E. Smalley, and C. Dekker, 'Electronic structure of atomically resolved carbon nanotubes', *Nature*, **391**, 59 (1998).

(3.51) T. W. Odom, J.-L. Huang, P. Kim and C. M. Lieber, 'Atomic structure and electronic properties of single-walled carbon nanotubes', *Nature*, **391**, 62 (1998).

(3.52) J. Liu, H. Dai, J. H. Hafner, J. T. Colbert, R. E. Smalley, S. J. Tans and C. Dekker, 'Fullerene "crop circles"', *Nature*, **385**, 780 (1997).

(3.53) R. C. Haddon, 'Electronic properties of carbon toroids', *Nature*, **388**, 31 (1997).

(3.54) M. F. Lin and D. S. Chuu, 'Electronic states of toroidal carbon nanotubes', *J. Phys. Soc. Japan*, **67**, 259 (1998).

(3.55) P. Nikolaev, A. Thess, A. G. Rinzler, D. T. Colbert and R. E. Smalley, 'Diameter doubling of single-wall nanotubes', *Chem. Phys. Lett.*, **266**, 422 (1997).

(3.56) Y. Chai, T. Guo, C. M. Jin, R. E. Haufler, L. P. F. Chibante, J. Fure, L. H. Wang, J. M. Alford and R. E. Smalley, 'Fullerenes with metals inside', *J. Phys. Chem.*, **95**, 7564 (1991).

(3.57) C. Yeretzian, K. Hansen, F. Diederich and R. L. Whetten, 'Coalescence reactions of fullerenes', *Nature*, **359**, 44 (1992).

(3.58) I. Mochida, M. Egashira, Y. Korai and K. Yokogawa, 'Structural changes of fullerene by heat treatment up to graphitization temperature', *Carbon*, **35**, 1707 (1997).

(3.59) Y. Ando and S. Iijima, 'Preparation of carbon nanotubes by arc-discharge evaporation', *Jap. J. Appl. Phys.*, **32**, L107 (1993).

(3.60) H. Terrones and M. Terrones, 'The transformation of polyhedral particles into graphitic onions', *J. Phys. Chem. Solids*, **58**, 1789 (1997).

(3.61) M. Ge and K. Sattler, 'Observation of fullerene cones', *Chem. Phys. Lett.*, **220**, 192 (1994).

(3.62) A. Krishnan, E. Dujardin, M. M. J. Treacy, J. Hugdahl, S. Lynum and T. W. Ebbesen, 'Graphitic cones and the nucleation of curved carbon surfaces', *Nature*, **388**, 451 (1997).

(3.63) M. Kosaka, T. W. Ebbesen, H. Hiura and K. Tanigaki, 'Annealing effect on carbon nanotubes. An ESR study', *Chem. Phys. Lett.*, **233**, 47 (1995).

4

The physics of nanotubes

When we get to the very, very small world – say circuits of seven
atoms – we have a lot of new things that would happen that
represent completely new opportunities for design. Atoms on a
small scale behave like *nothing* on a large scale, for they satisfy the
laws of quantum mechanics.

Richard Feynman, *There's plenty of room at the bottom*

With diameters frequently less than 10 nm, carbon nanotubes fall into the size
range where quantum effects become important, and this, combined with their
unusual symmetries, has led theoreticians to predict some remarkable electronic, magnetic and lattice properties. As noted in Chapter 1, several groups
around the world showed theoretically in late 1991 that nanotubes can be
metals or semiconductors depending on their precise structure and diameter.
This raised intriguing possibilities such as the idea of creating 'insulated'
nanowires, in which a conducting tube would be isolated inside a semiconducting one. Other workers pointed out that nanotubes might exhibit exotic
quantum mechanical behaviour such as the Aharonov–Bohm effect in the
presence of a magnetic field. Initially, testing the theoretical predictions proved
problematic. Measurements on the bulk material were difficult to interpret
because bulk samples inevitably contained a wide variety of tube structures and
sizes, as well as other material such as nanoparticles. The development of
methods for purifying nanotube samples enabled some progress to be made, but
the results were still rather difficult to interpret. Recently, however, a number of
groups have succeeded in carrying out electronic measurements on individual
nanotubes, and in constructing nanotube-based electronic devices. These remarkable experiments are among the highlights of nanotube research to date.

The aim of this chapter is to review the theoretical and experimental work
on the physical properties of carbon nanotubes, concentrating primarily on
their electronic, magnetic and vibrational properties, since these are the areas

that have been most widely studied. To begin with, a brief summary of the electronic structure and transport properties of graphite and carbon fibres is given.

4.1 Electronic properties of graphite and carbon fibres

4.1.1 Band structure of graphite

As one would expect from its structure, the electronic properties of graphite are highly anisotropic. Electron mobility within the planes is high, as a result of overlap between the π orbitals on adjacent atoms, and the room temperature in-plane resistivity of high-quality single-crystal graphite is approximately $0.4\,\mu\Omega\,\mathrm{m}$. However, mobility perpendicular to the planes is relatively low. The first detailed band structure calculations for graphite, by P. R. Wallace in 1947 (4.1), were carried out for conduction solely in the planes, and ignored any interactions between planes. The following expression for the energy, E_{2D}, of an electron at a point defined by the wavevectors k_x, k_y was obtained:

$$E_{2D}(k_x,k_y) = \pm \gamma_0 \left\{ 1 + 4\cos\left(\frac{\sqrt{3}k_x a}{2}\right)\cos\left(\frac{k_y a}{2}\right) + 4\cos^2\left(\frac{k_y a}{2}\right) \right\}^{1/2} \quad (4.1)$$

where γ_0 is the nearest-neighbour transfer integral and $a = 0.246\,\mathrm{nm}$ is the in-plane lattice constant.

The unit cell for 2D graphite contains two atoms, so we have four valence bands, three σ and one π. The above expression produces bonding and antibonding π bands which just touch at the corners of the hexagonal 2D Brillouin zone, so that at zero kelvin the bonding π band would be completely full and the antibonding π band completely empty. Figure 4.1(a) shows the E vs. k curves for 2D graphite along the direction K–Γ–M in the Brillouin zone; a sketch of the Brillouin zone for 2D graphite is given in Fig. 4.1(b). The density of states near the Fermi level for 2D graphite is shown in Fig. 4.2(a).

The band structure of three-dimensional graphite was calculated by Slonczewski, Weiss and McClure in the 1950s (4.3–4.5). The model shows that the bands overlap by $\sim 40\,\mathrm{mev}$, making graphite a semi-metal with free electrons and holes at all temperatures. A sketch of the density of states near the Fermi level for 3D graphite is shown in Fig. 4.2(b). A detailed discussion of the band structure of graphite is not necessary here, but excellent reviews have been given by a number of authors (4.2, 4.5–4.8).

The Slonczewski–Weiss–McClure (SWMcC) model allows us to calculate the electronic transport properties of graphite. However, such calculations are difficult, and a close agreement with experiment is not always obtained (4.2).

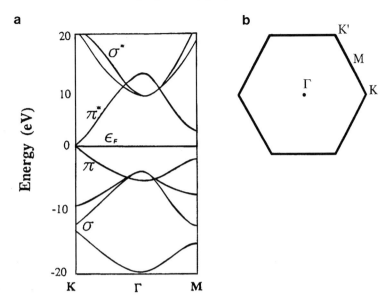

Fig. 4.1. (a) Dispersion relation for 2D graphite along the directions K–Γ and Γ–M in the Brillouin zone. (b) Sketch of the Brillouin zone for 2D graphite.

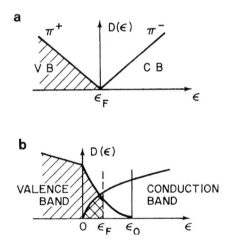

Fig. 4.2. Density of states near the Fermi level for (a) 2D graphite, (b) 3D (Bernal) graphite (4.2).

The calculations show that graphite has a carrier density of the order of 10^{18} cm^{-3}, i.e. about one carrier per 10^4 atoms. Thus the conductivity will be very low compared with, say, copper which has one free carrier per atom. The low carrier density is partly offset by relatively high carrier mobilities in the basal plane, and this is reflected in the temperature dependence of resistivity.

4.1.2 Transport properties of graphite, disordered carbons and carbon fibres

The electrical properties of graphites and graphitic carbons can vary enormously, depending on the degree of crystalline order. Experimental measurements of the variation in basal plane resistivity with temperature for a range of carbons are shown in Fig. 4.3. In the case of high-quality single-crystal graphite, the resistivity falls quite rapidly with falling temperature, particularly in the range below about 100 K (curve (1)). The fall in resistivity is chiefly due to a reduction in phonon scattering at low temperatures. On the other hand, more disordered carbons can exhibit a rise in resistivity as temperature falls. This is a consequence of falling carrier density at low temperatures, which is the dominant factor in determining resistivity in such carbons.

It should be noted that the electronic transport properties of graphite can be radically transformed by intercalation (4.9). When intercalated with acids such as AsF_5, room temperature conductivity as high as that of copper has been observed, while doping with potassium produces the superconducting intercalate C_8K.

As far as carbon fibres are concerned, experimental studies show that these also vary quite widely in their electronic properties (4.10). Thus, fibres of the highest perfection have resistivity curves approaching that of single-crystal graphite while less perfect fibres exhibit behaviour similar to that of disordered carbons.

4.1.3 Magnetoresistance of graphite and carbon fibres

The electronic behaviour of graphitic materials in the presence of a magnetic field can provide valuable information about carrier mobility and about crystalline order, and has been widely studied. For crystalline conductors, the effect of a magnetic field is normally to increase the resistivity, that is they exhibit positive magnetoresistance. Graphitic carbons are unusual in that they can exhibit positive or negative magnetoresistance. For high-quality single-crystal graphite, such as HOPG, positive magnetoresistance is invariably observed. Turbostratic carbons, on the other hand, tend to exhibit negative magnetoresistance, indicating that the degree of 3D interplanar ordering is the most important factor determining magnetoresistance behaviour. Disordered carbons also show negative magnetoresistance. Carbon fibres can display either positive or negative magnetoresistance, depending on the degree of graphitization. For example, benzene derived vapour-grown fibres heat treated at 2200 °C and above exhibit positive magnetoresistance at liquid nitrogen temperatures, while those heated at lower temperatures show nega-

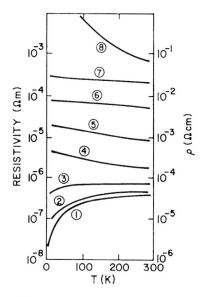

Fig. 4.3. The temperature dependence of resistivity for various forms of carbon. (1) Single crystal graphite, (2) highly oriented pyrolytic graphite, (3) graphite whisker, (4) pyrolytic carbon, (5) petroleum coke carbon, (6) lampblack base carbon, (7) glassy carbon, (8) evaporated carbon film (4.2).

tive magnetoresistance (4.11). A detailed review of magnetoresistance in graphitic materials has been given by Delhaes (4.12).

4.2 Electronic properties of nanotubes: theory

4.2.1 Band structure of single-walled tubes

In determining the band structure of graphite, it is assumed that the graphene planes are infinite in two directions, and artificial boundary conditions are introduced on a macroscopic scale in order to carry out the calculation. For carbon nanotubes, we have a structure which is macroscopic along the fibre axis, but with a circumference of atomic dimensions. Therefore, while the number of allowed electron states in the axial direction will be large, the number of states in the circumferential direction will be very limited. The allowed states can be thought of as lying on a number of parallel lines within the 2D graphene Brillouin zone. The nanotube BZ is then constructed by 'compressing' these lines into a single line. This will now be discussed in more detail, drawing on the work of Mildred Dresselhaus and co-workers from MIT, and Noriaki Hamada and colleagues from Iijima's laboratory in Tsukuba (4.13–4.16) (although, as noted in Chapter 1, it is recognised that the first electronic structure calculations for carbon nanotubes were actually carried out by a group from the Naval Research Laboratory in Washington (4.17)).

We consider armchair tubes first. Using the well-known expression for a periodic boundary condition (4.18), allowed values for the wavevector in the circumferential direction can be written as,

$$k_x^\nu = \frac{\nu}{N_x} \frac{2\pi}{\sqrt{3}a} \tag{4.2}$$

for $\nu = 1, \ldots, N_x$.

Taking the example of the 'archetypal' (5,5) armchair tube, ν has the values 1, . . ., 5. Thus, there are five allowed modes in the x direction in this case, so that the one-dimensional energy dispersion relations lie along five lines on either side of the centre of the Brillouin zone with a further line passing through the centre, as shown in Fig. 4.4. Now, it was noted above that the valence and conduction bands for graphite are degenerate at the K point. Therefore, nanotubes with a set of wavevectors which include the K point should be metallic. For armchair tubes, it is clear from Fig. 4.4 that the orientation of the Brillouin zone means that there will always be one set of allowed vectors passing through the K point, which leads to the conclusion that all armchair tubes are metallic. This remains true even though the Fermi wavevector is displaced slightly from the ideal K point as a result of the curvature of the tube.

The energy dispersion relation for a (5,5) armchair tube is obtained by substituting the allowed values of k_x^ν into Equation (4.1), and is shown in Fig. 4.5(a). Each band can be assigned to an irreducible representation of the D_{5d} point group, and is labelled accordingly in the figure. The A bands are non-degenerate and the E bands are doubly degenerate, so the total number of valence bands in this case is 10; the + and − labels denote the unfolded and folded bands respectively. It can be seen that the valence and conduction bands touch at a position which is two-thirds of the distance from $k = 0$ to the zone boundary at $k = \pi/a$. Calculations show that all armchair tubes have a similar band structure.

For zig-zag tubes the allowed wavevectors are given by

$$k_y^\nu = \frac{\nu}{N_y} \frac{2\pi}{a} \tag{4.3}$$

for $\nu = 1, \ldots, N_y$.

Thus, for the (9,0) tube there are nine lines of allowed wavevectors, as shown in Fig. 4.6. The energy dispersion relation for this case is shown in Fig. 4.5(b),

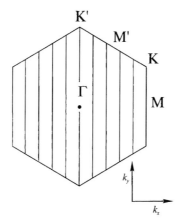

Fig. 4.4. Illustration of allowed k values in Brillouin zone for (5,5) armchair nanotube.

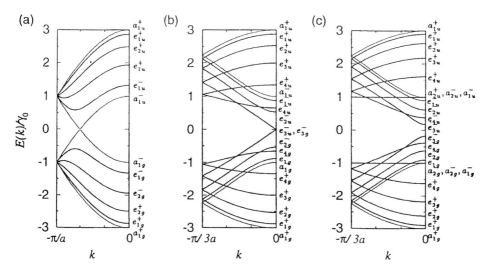

Fig. 4.5. Dispersion relations for (a) armchair (5,5) nanotube, (b) zig-zag (9,0) nanotube, (c) zig-zag (10,0) nanotube (4.15).

where the bands are assigned to an irreducible representation of the D_{9d} point group. Here, there are two non-degenerate A bands and eight doubly degenerate E bands making a total of 18. The valence and conduction bands touch at $k = 0$, so that in this case the tube is a metal. The reason for this is clear from Fig. 4.6, where it can be seen that one of the lines of allowed wavevectors for this tube passes through a K point. This is not the case for all zig-zag tubes, and only occurs when n is divisible by three. Thus, for a (10,0) tube there is an

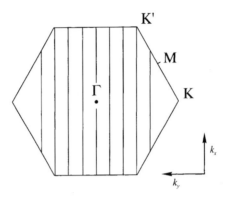

Fig. 4.6. Illustration of allowed k values in Brillouin zone for (9,0) zig-zag nanotube.

energy gap between the valence and conduction bands at $k = 0$, as shown in Fig. 4.5(c), and the tube would be expected to be a semiconductor. The electronic density of states for the (9,0) and (10,0) zig-zag tubes have been calculated by the Dresselhaus group (4.19), and are shown in Fig. 4.7. It can be seen that there is a finite density of states at the Fermi level for the metallic (9,0) tube, and a vanishing density of states for the semiconducting (10,0) tube.

Chiral nanotubes may also be either metallic or semiconducting, depending on chiral angle and tube diameter. Dresselhaus *et al.* (4.14, 4.15, 4.19) show that metallic conduction occurs when

$$n - m = 3q$$

where n and m are the integers which specify the tube's structure and q is an integer. Thus, one-third of chiral tubes are metallic and two-thirds are semiconducting.

In summarising the above discussion we can say that all armchair single-walled carbon nanotubes are expected to be metallic, while one-third of zig-zag and chiral tubes should be metallic, with the remainder being semiconducting.

Finally, when considering the electronic properties of one-dimensional conductors, it is important to consider the possibility of a Peierls distortion. This effect, predicted many years ago by Rudolf Peierls (4.20), involves a lattice distortion at low temperatures which results in the splitting of the uppermost band of a one-dimensional metal, thus transforming it into a semiconductor. Peierls instabilities are observed in one-dimensional molecular conductors, and have been a serious problem in the search for superconductivity in such systems. A number of authors have considered the possibility of a Peierls instability in carbon nanotubes (e.g. 4.17, 4.21), and the general conclusion

(a)

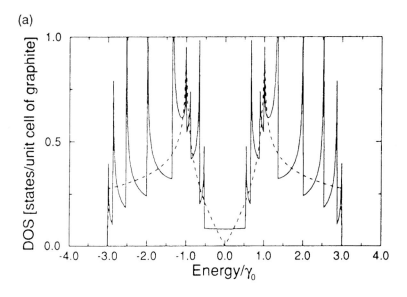

(b)

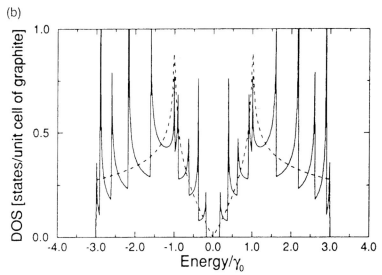

Fig. 4.7. Electronic 1D density of states per unit cell for two zig-zag tubes (4.19). (a) The (9,0) metallic tube, (b) the (10,0) semiconducting tube. The dotted line shows the density of states for a 2D graphene sheet.

seems to be that metallic tubes should be quite stable against such distortions. This appears to be borne out by some of the recent experimental work on single-walled tubes (see Section 4.3.2).

4.2.2 Band structure of multiwalled tubes

So far, only single-walled tubes have been considered. The electronic proper-ties of multiwalled tubes have also been discussed theoretically. Riichiro Saito and colleagues calculated the band structure of double-walled nanotubes using the tight-binding method (4.22). They showed that interlayer coupling had little effect on the electronic properties of the individual tubes. Thus, two coaxial zig-zag nanotubes that would be metallic as single-walled tubes yield a metallic double-walled tube. Semiconducting tubes behaved similarly. They also showed that coaxial metallic–semiconducting and semiconducting–me-tallic tubes retained their respective characters when interlayer interactions were introduced. Therefore, the idea that double-layer nanotubes could be used as insulated nanowires appears to be at least theoretically possible.

Subsequent work by a group from the Université Catholique de Louvain, Belgium (4.23, 4.24) has suggested that the situation may not be quite so simple. These workers considered the case of a (5,5) tube inside a (10,10) tube, both of which would be metallic as individual tubes. Unlike Saito *et al.*, they looked at the effect of changing the relative position of one tube with respect to the other and found that in certain configurations the interlayer interactions can cause both tubes to become semiconducting. On the other hand, when they considered a (10,10) tube inside a (15,15) tube, both tubes remained metallic. Further work is required to determine the electronic properties of multilayer zig-zag and chiral nanotubes.

4.2.3 Electron transport in nanotubes

We have seen in the above discussion that the allowed electronic states for carbon nanotubes will be very limited compared with those for bulk graphite. This is illustrated in Fig. 4.7, which shows the allowed states for two zig-zag tubes. The consequence of this is that the transport behaviour of nanotubes will be essentially that of a quantum wire, so that conduction occurs through well-separated, discrete electron states. Experimental work on single-walled tubes, discussed in Section 4.3.2, has appeared to show that the tubes do indeed display this behaviour. Another aspect of quantum wire behaviour is that transport along the tubes may be ballistic in nature. Ballistic transport occurs when electrons pass along a conductor without experiencing any scattering from impurities or phonons; effectively, the electrons encounter no resistance, and dissipate no energy in the conductor. There is currently great interest in exploiting this phenomenon to construct ultrafast devices (4.25), but fabricating structures with the desired properties has proved difficult. Ballistic conduction in nanotubes has been discussed theoretically (4.26, 4.27), and recent experimental work provides evidence that the phenomenon occurs in multi-walled tubes (see Section 4.3.1).

4.2.4 Nanotube junctions

It was mentioned in Chapter 3 (Sections 3.3.6 and 3.5.8) that 'elbow' connections between tubes of different structures are quite frequently observed in samples prepared by arc-evaporation. The electronic properties of such junctions have been considered theoretically (e.g. 4.28, 4.29). It has been pointed out that a connection between a metallic tube and a semiconducting one would represent a nanoscale heterojunction. Thus, higher energy electrons from the semiconducting side of the junction could flow 'downhill' to the metallic side, but they could not travel the other way. Such junctions are of course the basis for many types of electronic device. Recent work by Collins *et al.* (4.30) seems to provide evidence for just this kind of behaviour in single-walled tubes, as discussed in Section 4.3.2 below. However, the synthetic control required to produce such junctions at will is lacking at present.

4.2.5 Electronic properties of nanotubes in a magnetic field

The effect of a magnetic field on the electronic properties of graphite and of carbon fibres was briefly discussed in Section 4.1.3. The electronic properties of nanotubes in a magnetic field have been discussed by several authors, and this work will now be summarised.

Consider first a magnetic field applied parallel to the tube axis. Calculations by Hiroshi Ajiki and Tsuneya Ando of the University of Tokyo, using **k.p** perturbation theory (4.31, 4.32), indicate that in this case the band gap would oscillate with increasing magnetic field, so that a metallic tube would become firstly semiconducting and then metallic again, with a period dependent on the magnetic field strength. This behaviour is a consequence of the Aharonov–Bohm effect, which is another phenomenon characteristic of quantum wires. In the case of nanotubes, the Aharonov–Bohm effect means that a magnetic field changes the boundary conditions which determine how the 2D graphene energy bands are cut. The variation of energy gap with magnetic flux for an initially metallic nanotube is shown in Fig. 4.8 (4.31). Here, the magnetic

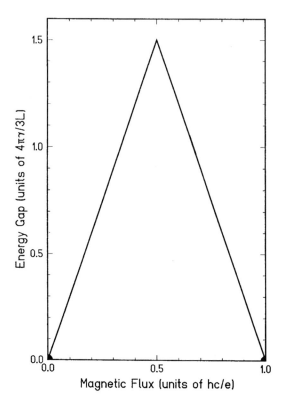

Fig. 4.8. The variation of energy gap with magnetic flux for an initially metallic nanotube (4.31).

flux is given in units of ϕ_0, the flux quantum, defined by $\phi_0 = hc/e$, where h is Planck's constant, c is the velocity of light and e is the electronic charge. The magnitude of the magnetic field required to deliver a flux quantum deceases rapidly with increasing nanotube diameter. Thus, for a tube with diameter 0.7 nm the magnetic field required is 10 700 T, while for a 30 nm tube the field would be 5.85 T. Since fields larger than about 30 T cannot be easily achieved, the Aharonov–Bohm oscillations predicted by Ajiki and Ando would probably only be observable in practice for relatively large diameter tubes.

Oscillations of electron energy band gap with increasing magnetic field are also predicted when the field is applied *perpendicular* to the tube axis. This case was also considered by Ajiki and Ando using the **k.p** method (4.31), and the tight-binding approximation (4.32). Figure 4.9 shows the band gap as a function of magnetic field for three zig-zag nanotubes with different circumferences. Here the energy gap is expressed in units of $\gamma_0(a/L)^2$ where γ_0 is the nearest-neighbour transfer integral, a is the unit cell base vector and L is the nanotube circumference, and the magnetic field is expressed in units of $L/2\pi l$ where

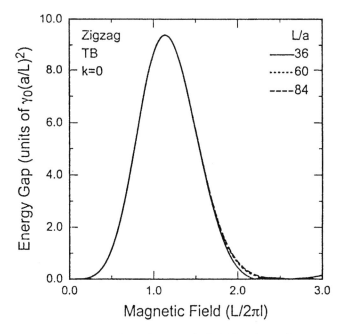

Fig. 4.9. The band gap as a function of magnetic field for three zig-zag nanotubes with different diameters (4.32).

$l = \sqrt{ch/eH}$ and H is the magnitude of the magnetic field. The three zig-zag tubes chosen would all be metallic in the absence of a magnetic field, and it can be seen that the variation of band gap with magnetic field is identical in each case. The field required to produce the maximum band gap for a (60,0) tube (diameter $= 4.7$ nm) would be 220 T. The electronic properties of nanotubes in a magnetic field perpendicular to the tube axis have also been considered by Saito and colleagues (4.33) with broadly similar results.

4.3 Electronic properties of nanotubes: experimental measurements

4.3.1 Resistivity measurements on multiwalled nanotubes

An early attempt to measure the resistivity of nanotube-containing samples was made by Ebbesen and Ajayan, following their success in developing a bulk synthesis method in 1992 (4.34). They found that raw cathodic soot had a resistivity of approximately 100 μΩ m. This figure is relatively high in comparison with the resistivities of conventional carbons shown in Fig. 4.3, probably as a result of contact resistances between tubes in the path of the current. Also, the unpurified material contains nanoparticles and disordered material, which would increase its resistivity. Some of the individual tubes would be expected to have a much lower resistivity than this bulk value. A number of subsequent studies were made on microscopic bundles of nanotubes and on purified samples, but these have been largely superseded by measurements on individual tubes, and will not be discussed here.

Workers from the Catholic University of Louvain in Belgium were among the first to report electrical measurements on individual multiwalled nanotubes (4.35). Their method involved dispersing nanotube 'bundles' on an oxidised silicon wafer onto which an array of gold pads had been deposited, and then evaporating a further gold film over the whole wafer. STM was then used to identify bundles situated between two pads, and photolithographic methods were used to make connections between the bundles and the pads. The arrangement is illustrated in Fig. 4.10. Results were reported for a particular nanotube which had a diameter of $\sim 20\,\text{nm}$, with a distance of $\sim 800\,\text{nm}$ between the contacts. Electrical resistance was determined as a function of temperature, down to $T = 30\,\text{mK}$; and as a function of magnetic field. Resistance was found to rise with falling temperature, being approximately proportional to $-\ln T$ for temperatures above about 1 K. Below this temperature, the resistance reached saturation at around 0.01 K. The effect of a magnetic field applied perpendicular to the tube axis was to reduce the resistance at all temperatures, so that the tube displayed negative magnetoresistance. The authors interpret the $\ln T$ resistance behaviour in terms of weak localisation, i.e. confinement of electrons in the region of individual atomic sites. This kind of behaviour is not unusual for disordered carbons. Langer and colleagues also revealed fluctuations in the resistance as a function of magnetic field at low temperature, as shown in Fig. 4.11. They ascribe this behaviour to the Aharonov–Bohm effect, as predicted by Ajiki and Ando (4.31, 4.32).

A short time later Charles Lieber and colleagues from Harvard University also described conductivity measurements on individual nanotubes (4.36). They used a slightly different method to that of the Belgian group, which involved firstly depositing the nanotubes onto an oxidised silicon wafer and coating with a layer of gold, into which a pattern of slots was then made. This left many single nanotubes extending into an open region, as shown in Fig. 4.12. Individual nanotubes could be located using atomic force microscopy, and conductivity measurements on the tubes made using a conducting cantilever tip assembly. An advantage of this method over that of Langer and colleagues is that resistance measurements can be made at several points along a given tube, thus eliminating contact resistance. The Harvard group used nanotubes prepared catalytically rather than by arc-evaporation for their study, since this method produces single tubes rather than clusters. A disad-

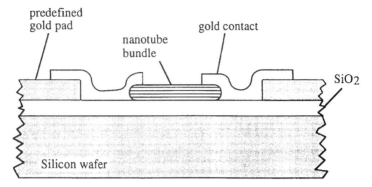

Fig. 4.10. Method used by Langer *et al.* to make electrical connections to nanotubes (4.35). Although a nanotube bundle is shown here, the same arrangement was used to make connections to individual tubes.

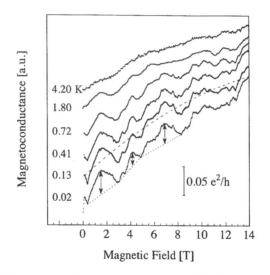

Fig. 4.11. Magnetoconductance measurements by Langer *et al.* (4.35).

vantage is that catalytically formed tubes are usually less perfect than those produced by arc-evaporation.

Measurements were made on six tubes, two straight and four curved. The straight tubes had diameters of 8.5 nm and 13.9 nm, and were found to have resistances of 0.41 MΩ/μm and 0.06 MΩ/μm respectively. In order to compare the electrical properties of the two tubes independent of their size, Lieber and colleagues calculated their resistivities, using estimates of their annular cross-sectional areas. They found that the 8.5 nm tube had a resistivity of 19.5 μΩ m and the 13.9 nm tube had a resistivity of 7.8 μΩ m. These figures are

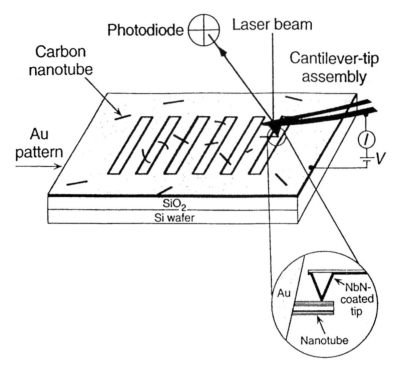

Fig. 4.12. The arrangement used by Lieber *et al.* to measure resistivities of individual nanotubes (4.36).

in reasonable agreement with those of Langer and co-workers. Although Langer *et al.* did not carry out measurements at room temperature, their results can be extrapolated to give a room temperature value of the resistivity. Thus, the room temperature resistance of their 20 nm diameter nanotube would be approximately $34\,k\Omega/\mu m$, implying a resistivity in the region of $8\,\mu\Omega\,m$. Taken together, these three figures appear to indicate that resistivity falls with increasing diameter, although many more measurements will be needed to confirm this. Lieber and colleagues also measured the resistivities of bent nanotubes and found them to be higher than those of straight tubes.

It is notable that the values of resistivity for individual tubes obtained by Langer *et al.* and by Lieber and colleagues are an order of magnitude lower than the value obtained in the study of unpurified nanotubes by Ebbesen and Ajayan. However, the values are still much higher than the room temperature in-plane resistivity of high-quality graphite, which is approximately $0.4\,\mu\Omega\,m$.

The most comprehensive study to date of the electrical properties of individual multiwalled nanotubes was reported in mid-1996 by Thomas Ebbesen and colleagues of NEC, in collaboration with Micrion Europe (4.37). Before carrying out electrical measurements, these workers annealed the nanotubes at 2850 °C, since ESR measurements had demonstrated that such a treatment can remove defects (see Section 4.3.4). The tubes were then deposited onto an oxidised silicon surface between gold pads. A focused ion beam microscope was used to image the supported nanotubes, and when a suitable tube had been located, four 80 nm-wide tungsten wires were deposited to produce an arrangement such as the one illustrated in Fig. 4.13. The tungsten leads could then be connected to gold pads to enable four-probe resistance measurements to be carried out. The distance between contacts on the tubes was in the range 0.3–1.0 μm. In order to measure the temperature effects on resistance, the samples were mounted on a cryostat.

Resistance measurements were reported for eight different nanotubes and the results differed widely. The highest measured resistance, for a tube with a diameter of 10 nm was greater than $10^8 \, \Omega$, while the lowest resistance, for an 18.2 nm tube, was $2 \times 10^2 \, \Omega$. In both cases the distance between leads onto the tubes was 1.0 μm. Ebbesen and colleagues estimated that these figures translate to resistivity values of 8 mΩ m and 0.051 μΩ m respectively. Although these values are only very approximate, they show that the room temperature resistivity of nanotubes can in some cases be comparable to, or lower than, the in-plane resistivity of graphite.

The temperature dependence of resistivity of the tubes also differed widely, as shown in Fig. 4.14. In addition to the variation from tube to tube, different segments of a single tube could sometimes have different temperature profiles. This can be seen in Fig. 4.14(b) and Fig. 4.14(d), which show the curves for two different lengths of tube NT7. The commonest type of behaviour was that exhibited by tubes NT8, NT7b, NT2, NT1, namely a consistent slight increase in resistivity with decreasing temperature. Ebbesen and colleagues do not consider that these tubes should be thought of as semiconducting, particularly in view of their low resistivities. Instead they suggest that the tubes are essentially metallic, and the variations in their resistivities and temperature dependence are due to the interplay of changes in carrier concentration and mobilities. In other cases, however, tubes did display semiconducting behaviour. The temperature profile of tube NT7a is unusual, showing a sharp drop in resistance which flattens off as the temperature is lowered below 220 K. Ebbesen and colleagues suggest that this may indicate an insulator-to-metal transition. The behaviour of tube NT6 is anomalous, with resistance becoming unmeasurably large for temperatures below 200 K. This is attributed to a complex conductive path inside the tube. The authors point out that the conductivity behaviour of multilayer nanotubes is likely to be complicated by the intertube interactions and the presence of defects.

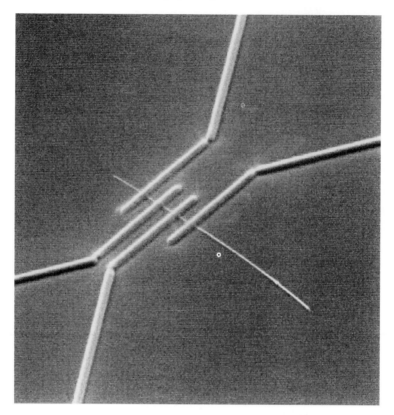

Fig. 4.13. Focused ion beam image of four tungsten wires connected to an individual nanotube, from the work of Ebbesen *et al.* (4.37). Each tungsten wire is 80 nm wide.

Ebbesen and co-workers also examined the effect of applying a magnetic field perpendicular to the long axis of the tubes, but only found a very small effect. For example, a field of 10 T produced a positive magnetoresistance of only $\sim 1\%$ at a temperature of 7 K.

To summarise the discussion so far, resistivity measurements on multiwalled nanotubes show wide variations in behaviour. Since samples of multiwalled tubes are known to be structurally heterogeneous, these results provide clear evidence that the electronic properties of carbon nanotubes can vary greatly according to their structure.

Finally in this section, we should mention some remarkable experiments which appear to demonstrate that electrons flow ballistically through multi-walled nanotubes, and that the conductance of the tubes is quantised. Ballistic transport occurs when electrons encounter no resistance in the conductor, as discussed in Section 4.2.3 above. This behaviour was observed in nanotubes by Walt de Heer and colleagues in mid-1998 (4.38). These workers devised an

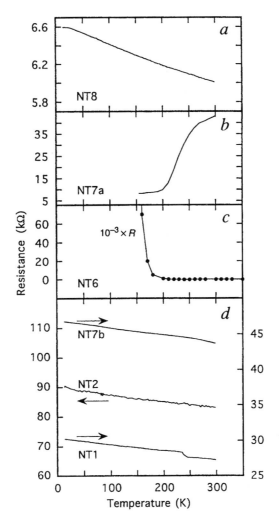

Fig. 4.14. The temperature dependence of resistance for different nanotubes, determined by Ebbesen *et al.* (4.37). (a)–(c) are four-probe measurements, while (d) shows two-probe results.

ingenious arrangement in which a nanotube bundle (at the extremity of which was a single tube) could be dipped into a heatable reservoir containing mercury, as shown in Fig. 4.15. When a circuit was established, the current I was measured as a function of the position of the nanotube in the mercury. It was found that the conductance did not change smoothly with position, as would be expected for a classical conductor, but instead increased in a series of jumps, consistent with conductance quantisation. The nanotubes were found to be undamaged, even at relatively high voltages (6 V) for extended times. De Heer *et al.* calculated that the power dissipation at these voltages would produce

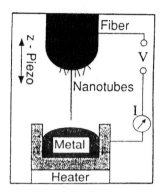

Fig. 4.15. Experimental setup used by de Heer *et al.* to measure resistance of individual multiwalled nanotubes (4.38).

enormously high temperatures (up to 20 000 K) in the tubes if they were acting as classical resistors. The survival of the nanotubes therefore provided strong evidence for ballistic transport.

4.3.2 Resistivity measurements on single-walled nanotubes

The preparation of so-called 'ropes' of single-walled nanotubes by Richard Smalley and colleagues has been described in Chapter 2. The electronic properties of these samples have now been probed, and some very striking results obtained (4.30, 4.39–4.45). The first measurements were reported in the *Science* paper which described the synthesis of nanotube ropes (4.39). Four-probe measurements were made, using multiwalled nanotube probes to determine the voltage drop across a short length of the rope. Resistivities in the range 0.34 $\mu\Omega$ m to 1.0 $\mu\Omega$ m were reported. These values are quite similar to those of Ebbesen and colleagues described in the previous section.

   An extension of these studies was described in a paper published in early 1997 (4.40). Here, measurements were made on 'mats' of nanotube ropes and on individual ropes, again using multiwalled nanotube bundles as voltage probes. Resistivities of the order of 60 $\mu\Omega$ m were observed for the 'as grown' mat; this could be reduced by a factor of three if the mats were pressed to improve rope–rope contacts. For individual ropes the resistivities ranged from 0.3 to 1.0 $\mu\Omega$ m, thus straddling the in-plane resistivity of graphite. The temperature dependence of resistivity of the rope samples was also determined in this study. For the mat samples, resistivity increased with temperature for T above about 200 K, but decreased with temperature for lower temperatures. For individual ropes similar behaviour was observed, but with a lower cross-over temperature of around 35 K. The metal-like behaviour for high T appeared to provide support for the view that the ropes consist largely of (10,10) tubes. The negative $d\rho/dt$ at low temperature is more difficult to explain, but may be a consequence of localisation.

A further study of the electronic properties of nanotube ropes was carried out by Smalley's group in collaboration with workers from the University of California at Berkeley (4.41). In this work, evidence was presented for single-electron transport in nanotube ropes. To carry out the measurements, ropes were dispersed on oxidised silicon wafers and connections were made using lithographically defined metal leads. A gate voltage could be applied to the chip carrier on which the sample was supported. The current–voltage charac-teristics were determined for a range of temperatures down to 1.3 K and a range of gate voltages. A series of $I–V$ curves for temperatures from 1.3 K to 290 K are shown in Fig. 4.16. At the higher temperatures the curves are almost linear, as expected, but below 10 K the curves display a flat region near $V = 0$. This suppressed conductance, or 'gap', is indicative of an effect known as Coulomb blockade which occurs when the bias voltage becomes lower than the energy needed to add a single electron to the tube. The authors investigated the effect of gate voltage on I/V characteristics at a temperature of 1.3 K. As shown in Fig. 4.17, the conductance at $V = 0$ went through a series of peaks as the gate voltage was changed. These results can be under-stood in terms of the effect of gate voltage on the energy levels of the tubes and the two connecting electrodes. At gate voltages corresponding to a Coulomb peak, the energy of the lowest empty state in the tube aligns with the electrochemical potential in the leads, and single electrons can tunnel on and off the tube. At gate voltages between peaks, tunnelling is suppressed because of a barrier equal to the single-electron charging energy. The authors believe these sharp peaks indicate that transport may have occurred along a single nanotube.

A similar study was carried out at about the same time by Smalley's group with workers from Delft University of Technology in the Netherlands (4.42). This time it was possible to isolate *individual* nanotubes, as illustrated in Fig. 4.18. Measurements were made at a temperature of 5 mK, and at a range of gate voltages. Again, gaps were observed in the current–voltage curves around zero bias voltage. As in the previous study, the gaps were suppressed for certain gate voltages and restored at others, providing clear evidence that single-electron transport occurred in the individual tubes. This was confirmed by experiments on the effect of a magnetic field on conductance behaviour.

The same group described a further study of single-walled nanotubes at the beginning of 1998 (4.43), in which the tubes were probed by a combination of STM and scanning tunnelling spectroscopy (STS). Direct atomic imaging of

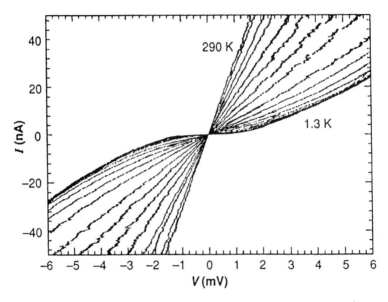

Fig. 4.16. *I–V* characteristics for nanotube rope segments at a range of temperatures (4.41).

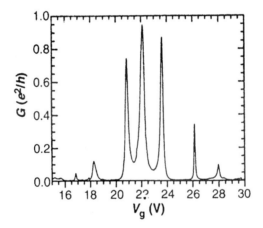

Fig. 4.17. Conductance *G* vs. gate voltage V_g for nanotube rope segment at $T = 1.3$ K (4.41).

the nanotubes by STM showed that a variety of nanotube structures were present, as previously discussed in Section 3.6.2. Most tubes were found to be chiral, with only a minority having armchair or zig-zag structures. The application of STS enabled the one-dimensional density of electronic states, given by the derivative of the *I–V* curve, to be determined along the tube axis. The chiral tubes were found to be either semiconducting or metallic, as predicted

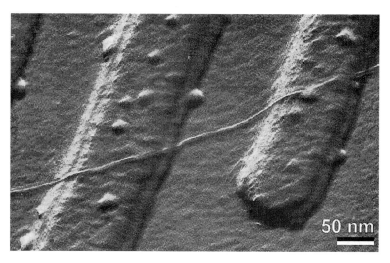

Fig. 4.18. Atomic force microscope image (tapping mode) of single-walled nanotube running between Pt electrodes on a Si/SiO$_2$ substrate (4.42).

by theory. The electronic properties of the armchair and zig-zag tubes also appeared to be consistent with theoretical predictions, although too few of these structures were probed to make any definitive statements. In the same issue of *Nature*, a group from Harvard described similar STM/STS studies of single-walled nanotubes produced by laser vaporisation (4.44). These two studies constitute the most detailed work yet performed on the electronic properties of nanotubes, and provide the clearest evidence so far that these properties depend on nanotube structure.

Two further papers on nanotube-based 'devices' have appeared more recently (4.45, 4.30). The Delft group have described a field-effect transistor consisting of a semiconducting single-walled nanotube connected to two metal electrodes (4.45). By applying a voltage to a gate electrode, the nanotube could be switched from a conducting to an insulating state. In contrast to their previous study (4.42), this behaviour was observed at room temperature. A slightly different kind of 'nanodevice' consisting solely of an individual single-walled nanotube has been described recently in a collaboration between the Smalley and Zettl groups and Hiroshi Bando of the Electrotechnical Laboratory, Tsukuba, Japan (4.30). In this work, an STM tip was moved along the length of individual tubes, and positions were found where the current transport behaviour changed abruptly. Effectively, the tubes passed current in only one direction. This may have resulted from the presence of pentagon–heptagon defects such as elbow connections. As discussed in 4.2.4, theoreticians have predicted that such defects should have this effect.

4.3.3 Doping of nanotube bundles

As noted in Section 4.1.2, doping of graphite with potassium produces a superconducting intercalate, while doping of solid C_{60} with alkali metals can result in superconducting transition temperatures in excess of 30 K (4.15). Therefore, doped bundles of single-walled nanotubes might be expected to have interesting electronic properties. Some initial conductivity measurements on doped nanotube ropes were described in mid-1997 (4.46, 4.47). The experiments employed similar methods to those used in the first measurements of superconductivity in doped C_{60} (4.48), and doping with potassium and bromine was investigated. In both cases, significant reductions in resistivity were observed, by up to a factor of 30. The region where the temperature coefficient of resistance was positive (indicating metallic behaviour) was also enlarged. These results were interpreted in terms of addition (for K) or removal (for Br) of electron density from the host carbon framework.

In a companion paper (4.47), Raman scattering experiments were reported on nanotube ropes doped with potassium, rubidium, iodine and bromine. These provided further evidence for charge transfer between the dopants and the nanotubes.

4.3.4 Electron spin resonance

Electron spin resonance (ESR) spectroscopy is a valuable probe of the electronic properties of solids and has been widely applied to graphite and other carbons. Measurements are made by placing a sample in a magnetic field (usually about 0.3 T), and observing the absorption of microwave radiation as the field is slowly varied. Information about the electronic properties of the sample can then be obtained by analysing the absorption intensity, line shape and g-shift, i.e the deviation of g from the free electron value of 2.0023. A number of ESR studies of carbon nanotubes have been made, and these will be discussed in detail below. Firstly, though, a brief discussion will be given of the application of the technique to single-crystal graphite and carbon fibres.

In pure graphite, the resonance is due solely to conduction carriers (4.49), and g-values are therefore highly anisotropic. When determined with the magnetic field parallel to the a–b planes, g is equal to 2.0026 ± 0.0002, close to the free electron value of 2.0023, while the value with the field perpendicular to the planes is 0.047 greater. The ESR lineshape for single-crystal graphite is characteristic of metal-like conduction (see Fig. 4.19). This shape is known as dysonian (4.50), and corresponds to the 'slow diffusion' case, in which the spins diffuse in and out of a microwave skin depth in a time which is long compared

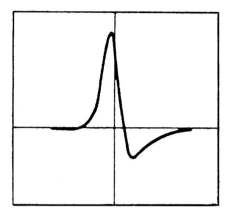

Fig. 4.19. Electron spin resonance spectrum of single crystal graphite showing dysonian lineshape (4.49).

to the spin–lattice relaxation time. For less highly ordered carbons, the analysis of ESR spectra is complicated because a large number of localised spins contribute to the absorption, in addition to the conduction carriers. Nevertheless, ESR has been widely used to study the electronic and structural properties of partially graphitized and disordered carbons (4.51).

Carbon fibres have also been quite extensively studied using ESR, and the technique has proved to be a useful indicator of crystalline perfection (e.g. 4.52). For all types of fibre it is found that the anisotropy of g, i.e. the difference between g-values recorded perpendicular to the fibres and those recorded parallel to the fibres, increases strongly with increasing heat-treatment temperature. For carbon fibres heated at the highest temperatures ($\sim 3000\,°C$), the lineshape is found to be dysonian.

There have been several ESR studies of carbon nanotubes, the most detailed of which was carried out by Kosaka, Ebbesen and colleagues (4.53, 4.54). These workers studied multiwalled nanotubes prepared by arc-evaporation and found that in most cases the lineshapes were approximately dysonian, indicating that a high proportion of the tubes were metallic (or had small band gaps). They also found that the ESR signals could be significantly affected by both purification and annealing. Figure 4.20 shows a set of ESR spectra taken from this work. In Fig. 4.20(a), the signals at 296 K and 10 K for crude nanotubes are shown. At room temperature a single peak, labelled A′, is observed, with a g-value of 2.018, while at 10 K there are three peaks, with the g-value of the A′ peak shifted to 2.043. Spectra for a sample of purified tubes are shown in Fig. 4.20(b). Here, the g-value of peak A is 2.012 at 296 K, and there are two peaks at 10 K. Kosaka *et al.* attributed the A peak to conduction electrons, and the

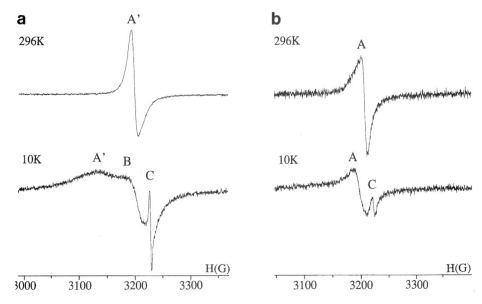

Fig. 4.20. ESR spectra at 296 K and 10 K for (a) crude nanotube sample and (b) purified sample (4.53).

low-temperature peaks to defects and impurities. The low-temperature peaks became less prominent following purification (Fig. 4.20(b)), and disappeared completely following annealing, as shown in Fig. 4.21. This was attributed to the removal of defects and the closure of opened tubes. These results are important, as they appear to show that the 'as-prepared' nanotubes are highly defective, and that a high proportion of the defects can be removed by annealing at high temperatures. Ebbesen and colleagues have suggested that, ideally, experimental work on the electronic and other properties of multiwalled nanotubes should be carried out on annealed samples, in order to eliminate the effect of defects.

Some ESR measurements on *aligned* multiwalled nanotubes were reported by Chauvet *et al.* of the Ecole Polytechnique Fédérale de Lausanne (4.55). Their alignment technique was described in Section 2.9, and involves firstly purifying the tubes and then coating them onto a Teflon supporting film. The spectra were recorded for a range of orientations at temperatures from 4 to 300 K. They found a single line over the entire temperature range and found a Pauli susceptibility of 10^{18} spins/cm^3, which is consistent with a semi-metal. They found relatively little variation in g with changing orientation. Thus, the value when the tubes were parallel to the field, $g_\parallel$, was found to be 2.0137, while $g_\perp$ was 2.0103. Unlike Ebbesen and colleagues, this group found the g-values to be temperature dependent.

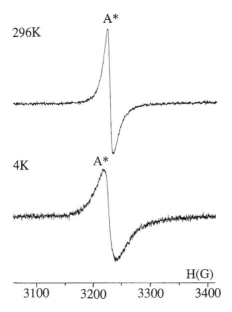

Fig. 4.21. ESR spectra for annealed purified nanotube sample (4.54).

Smalley and colleagues included ESR measurements in their first paper on nanotube ropes (4.39). Before carrying out the experiments, the samples were annealed to remove the ferromagnetic catalyst residues. The resulting spectra contained a narrow dysonian line at $g = 2.001$, consistent with a metal for which the skin depth is less than the sample dimensions. Subsequent ESR studies of 'rope' single-walled nanotubes (4.56) showed that the lineshape remained dysonian at temperatures down to 4 K. A spectrum taken from this work, and recorded at 100 K is shown in Fig. 4.22.

4.4 Magnetic properties of nanotubes

The magnetic properties of graphite are dominated by the presence of ring currents, i.e. electron orbits circulating above and below the hexagonal lattice planes which include several atoms within their radius (4.57). These result in a relatively large negative susceptibility which is highly anisotropic. Thus, when the field is oriented perpendicular to the layer planes the susceptibility, χ_c, is 22×10^{-6} emu/g, while the susceptibility with the field parallel to the planes, χ_{ab} is 0.5×10^{-6} emu/g. On a very simple model, one might assume that the magnetic properties of a carbon nanotube would approximate to those of a rolled-up graphene sheet. The susceptibility of tubes aligned perpendicular to the field, $\chi_\perp$, would therefore approximate to $(\chi_c + \chi_{ab})/2$ and the susceptibility of tubes aligned parallel to the field, $\chi_\parallel$,

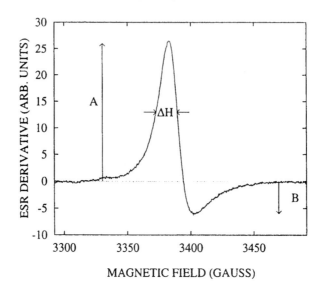

Fig. 4.22. ESR spectrum of single-walled nanotubes recorded at 100 K (4.56).

would approximate to χ_{ab}. Since $\chi_c \gg \chi_{ab}$, this would suggest that $\chi_\perp \gg \chi_\parallel$ for nanotubes. Detailed theoretical work has tended to confirm this simple-minded model (4.58–4.60).

Some early studies of the magnetic susceptibility of nanotubes were carried out by Arthur Ramirez and colleagues from Bell Labs, in collaboration with Smalley's group from Rice University (4.61). These workers studied a variety of carbons, including an unpurified sample of nanotubes, over a range of temperatures from absolute zero to room temperature, using a SQUID magnetometer. The results are shown in Fig. 4.23. It can be seen that the nanotube sample displays quite different behaviour to the other carbons, having a large diamagnetic susceptibility (i.e. negative χ) which increases with decreasing temperature. The results clearly indicate that nanotubes have a greater susceptibility than graphite, although it is not possible to say whether this susceptibility lies parallel or perpendicular to the tube axis since the tubes in the sample were randomly oriented. Ramirez *et al.* speculated that the large susceptibility of nanotubes might result from ring currents flowing around the tube circumferences.

The first attempts to measure the magnetic properties of aligned nanotube samples were made by X. K. Wang and colleagues from Northwestern University (4.62), who carried out their measurements on the columnar deposits which are sometimes produced on the cathode following arc-evaporation. Wang *et al.* have suggested that these deposits consist of aligned bundles of

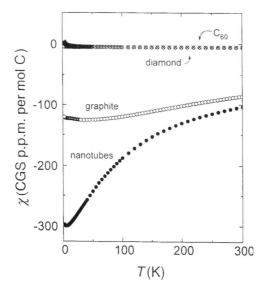

Fig. 4.23. Magnetic susceptibilities of carbon allotropes as a function of temperature, from work by Ramirez and colleagues (4.61).

nanotubes. Magnetic measurements on this material produced rather similar results to those of Ramirez *et al.*, but also revealed the presence of a small degree of anisotropy. It was found that the magnetic susceptibility with H parallel to the axis of the bundle was about $1.1 \times$ that of the susceptibility with H perpendicular to the axis, and that this anisotropy increased with decreasing temperature. These results represent the first evidence that carbon nanotubes have anisotropic magnetic properties, but it should be borne in mind that the degree of alignment within the bundles is probably not high.

Magnetic measurements on samples of nanotubes with a much higher degree of alignment were carried out by the Lausanne group (4.55). The ESR measurements on these samples were described in Section 4.3.4. For the magnetic measurements they employed films in which the tubes were aligned perpendicular to the film. Susceptibilities were measured from 4 to 300 K, with the tubes aligned parallel and perpendicular to the field, again using a SQUID magnetometer. The results confirmed that nanotubes are diamagnetic, and showed a pronounced anisotropy of susceptibility. However, as illustrated in Fig. 4.24, the magnetic susceptibility of tubes aligned parallel to the field ($\chi_{\parallel}$) was found to be much greater than that of tubes perpendicular to the field ($\chi_{\perp}$). This is the reverse of the theoretical predictions noted above, and the reason for this discrepancy is not clear.

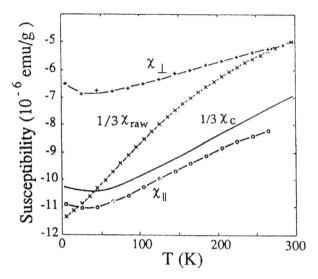

Fig. 4.24. Magnetic susceptibility vs. temperature for various nanotube samples, from the work of Chauvet *et al.* (4.55). Nanotubes with magnetic field perpendicular to tube axis (+); nanotubes with field parallel to tube axis (○); unprocessed nanotube-containing material (×); planar graphite (—).

4.5 Optical properties of nanotubes

There has been relatively little work on the optical properties of carbon nanotubes. However, a number of theoretical studies have been carried out, mainly by Chinese workers (e.g. 4.63–4.67). F. Huaxiang *et al.* calculated absorption coefficient ($\varepsilon_2(\omega)$) spectra for zig-zag and armchair nanotubes (4.63). These were found to be quite different, particularly for the lower energy region where the peaks correspond to transitions between π bands. A number of workers have analysed the non-linear optical properties of carbon nanotubes (4.66, 4.67), and found that these also depended strongly on the diameter and symmetry of the tubes. Testing these predictions directly would be difficult at present since pure samples of nanotubes with a given structure are not available.

One of the few experimental studies of the optical properties of nanotubes was carried out by Walt de Heer and colleagues in 1995 (4.68). These workers used ellipsometry to determine the dielectric function, ε, for tubes aligned either parallel or perpendicular to a support film. The perpendicular nanotube films were found to be optically isotropic, while for parallel films the optical properties were dependent strongly on whether the light was polarised parallel or perpendicular to the tubes. When the light was polarised along the tubes, the shape of the dielectric function was similar to that of HOPG. This is perhaps not surprising since in this case the surface resembles that of planar graphite. With the light polarised perpendicular to the tube direction, the dielectric function could be interpreted as a mixture of the two dielectric functions of graphite parallel and perpendicular to the graphitic sheets. The dielectric function for tubes aligned perpendicular to the substrate was similar to that of glassy carbon. More experimental work on the optical properties of nanotubes would be of great value.

4.6 Vibrational properties of nanotubes

4.6.1 Symmetry of vibrational modes

The phonon modes for carbon nanotubes can be obtained using similar methods to those employed in calculating the electronic properties. The symmetry assignments and spectroscopic activities of the vibrational modes for nanotubes have been discussed by Dresselhaus, Jishi, and co-workers (4.15, 4.69–4.72), and the following discussion is largely drawn from their work.

Since there are $2N$ atoms in the unit cell of a nanotube, the total number of vibrational modes will be $6N$. Group theory can be used to determine which of these modes are infra-red and Raman active. The modes can be decomposed into the irreducible representations of the point group appropriate to the tube. If we consider armchair tubes with D_{nh} symmetry (i.e. those for which n is an even integer), and assume that $n/2$ is odd, the vibrational modes will be decomposed into the following irreducible representations,

$$\Gamma^{vib} = 4A_{1g} + 2A_{1u} + 4A_{2g} + 2A_{2u} + 2B_{1g} + 4B_{1u} + 2B_{2g} + 4B_{2u} + 4E_{1g} + 8E_{1u} + 8E_{2g} + 4E_{2u} + \cdots + 8E_{(n/2-1)g} + 4E_{(n/2-1)u}.$$

If $n/2$ is even, the 8 and 4 in the last two terms are interchanged. Considering the (6,6) armchair tube, for which $N = 12$, this will have 72 phonon branches, and the above equation gives a total of 48 distinct mode frequencies (since the final two terms disappear). At the zone centre, one A_{2u} mode, one E_{1u} mode and one A_{2g} mode have zero frequencies since they correspond to translations along the tube axis or perpendicular to the axis, or to rotations about the axis.

Appropriate character tables can be used to determine which modes are IR active and which are Raman active. For the D_{nh} group, the A_{2u} and E_{1u} modes are IR active, while the A_{1g}, E_{1g} and E_{2g} modes are Raman active. Thus there are 8 IR mode frequencies ($A_{2u} + 7E_{1u}$) and 16 Raman active frequencies ($4A_{1g} + 4E_{1g} + 8E_{2g}$) at the zone centre. It is important to note that the number of IR and Raman active modes is independent of nanotube diameter;this is also true for all other classes of tube. However, the frequencies of

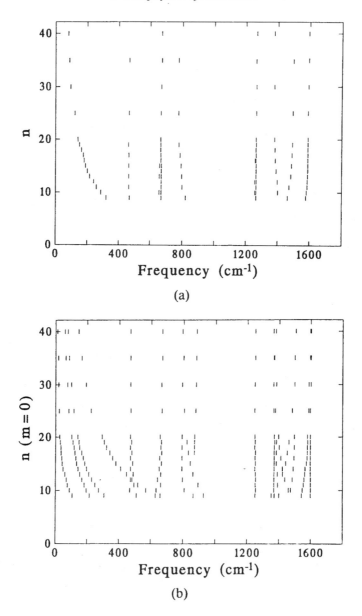

Fig. 4.25. Predicted diameter dependence of the first-order (a) IR active and (b) Raman active mode frequencies for zig-zag nanotubes (4.72).

these modes do vary with diameter, as discussed below.

In the case of armchair tubes with D_{nd} symmetry (n is odd), similar arguments show that there will be 7 IR active frequencies and 15 Raman active frequencies. Zig-zag tubes, whether D_{nh} or D_{nd} will also have 7 IR active frequencies and 15 Raman active frequencies.

Considering now chiral tubes, these belong to C_n-type symmetry groups, for which the vibrational modes contain the following symmetries,

$$\Gamma^{\mathrm{vib}} = 6A + 6B + 6E_1 + 6E_2 + \cdots + 6E_{N/2-1}.$$

In this case the A and E_1 modes are IR active and the A, E_1 and E_2 modes are Raman active. It should be noted that there is an important difference between armchair and zig-zag nanotubes on one hand and chiral tubes on the other. In the case of both armchair and zig-zag tubes, zone folding results in the M point being superimposed on the centre of the Brillouin zone, while for chiral tubes the M point does not map onto the Γ point. This suggests that there should be a larger spread of IR and Raman frequencies for armchair and zig-zag tubes than for chiral tubes.

The vibrational mode frequencies of nanotubes can be calculated from those of a two-dimensional graphene sheet using the equation

$$\omega_{1D}(k) = \omega_{2D}(k\hat{K}_2 + \mu\hat{K}_1),$$
$$\mu = 0,1,2, \ldots, N-1,$$

where ω_{1D} is the vibrational mode frequency for the one-dimensional tube, ω_{2D} is the frequency for a two-dimensional graphene sheet, k is a wavevector in the direction $\hat{K}_2$ in reciprocal space (i.e. along the tube axis) and μ is an integer used to label the wavevectors along the $\hat{K}_1$ direction in reciprocal space, perpendicular to the tube axis. Jishi, Dresselhaus and others have used this expression to calculate the variation of IR and Raman active modes with nanotube radius for a variety of different tube structures. The frequencies are calculated at the Brillouin zone centre, since only modes close to this point are Raman and IR active. As an example, Fig. 4.25 shows the diameter dependence of IR and Raman active mode frequencies for zig-zag tubes. The higher frequency modes can be seen to be less diameter-dependent than the lower frequency modes, and the diameter dependence becomes less marked for tubes with indices larger than about (20,0), corresponding to a diameter of 1.57 nm. For larger diameters the spectra tend towards those of graphite. The frequencies predicted for armchair and chiral nanotubes differ from those predicted for zig-zag tubes, suggesting that Raman and IR spectroscopy should provide a method of determining the chirality of nanotubes.

4.6.2 *Experimental IR and Raman spectra: multiwalled nanotubes*

The first Raman experiments on carbon nanotube-containing material were carried out by Thomas Ebbesen, Hidefumi Hiura and colleagues at NEC (4.73). These workers compared the Raman spectra of material from nanotubes and nanoparticles taken from the core of the cathode, with those from the outer shell of the cathodic deposit, and with the spectra of highly oriented pyrolytic graphite (HOPG) and of glassy carbon. The first order Raman spectra taken from their work are shown in Fig. 4.26. The strong peak which occurs in the region of $1580 \, \text{cm}^{-1}$ in each of the spectra can be assigned to one of the two Raman active E_{2g} vibrations of graphite, while the band at around $1350 \, \text{cm}^{-1}$ can be attributed to disorder. It can be seen that the spectrum from the nanotube-containing material and that from the outer shell are quite similar, and that both resemble the spectrum from HOPG rather more than that from glassy carbon. These demonstrate the high degree of crystallinity of the cathodic deposits compared to glassy carbon (in the case of the outer shell material, this is believed to consist largely of a fused mass of nanotubes and nanoparticles). However, the spectra do not give detailed information about nanotube structure since the core deposit will contain nanotubes with a variety of different sizes and structures, as well as nanoparticles.

Raman spectroscopy of multiwalled nanotubes was also described by Kastner and colleagues (4.74), with rather similar results to those of Hiura *et al.* These workers also recorded transmission IR spectra of nanotube-containing samples and found that the spectra resembled those from graphite microcrystallites, with a broad and unsymmetric line at $1575 \, \text{cm}^{-1}$ and a weaker line at $868 \, \text{cm}^{-1}$. Raman spectra of *purified* nanotube samples were reported by Bacsa and colleagues from the Ecole Polytechnique Fédérale de Lausanne in Switzerland (4.75). Purification was carried out by centrifugation followed by drying and annealing in air. In fact these workers found little difference between Raman spectra from purified and unpurified nanotube samples, both of which gave a narrow peak at $1581 \, \text{cm}^{-1}$. However, the 'disorder induced' line at $\sim 1350 \, \text{cm}^{-1}$ was found to be much smaller than in the other studies. The authors suggested that the strength of this feature may be related to the details of the arc-evaporation process.

4.6.3 *Experimental IR and Raman spectra: single-walled nanotubes*

The Raman and IR studies of multiwalled nanotubes discussed in the previous section produced results which did not differ greatly from those of single crystal graphite. Studies of single-walled nanotubes, on the other hand, have

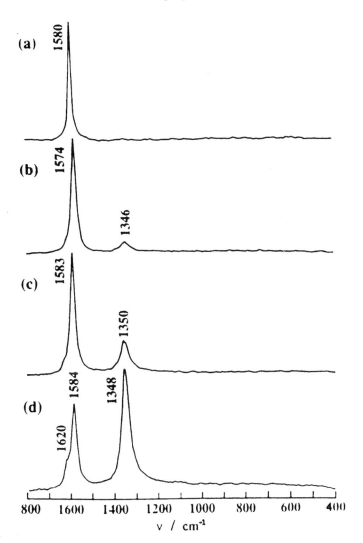

Fig. 4.26. Room temperature Raman spectra of (a) HOPG, (b) inner core of the cathodic deposit containing carbon nanotubes and nanoparticles, (c) outer shell of the cathodic deposit, (d) glassy carbon. From the work of Hiura *et al.* (4.73).

tended to produce significantly different spectra from that of graphite. One of the first such studies was reported by J. M. Holden and colleagues (4.76). Their samples were prepared using cobalt-containing electrodes, in the way described by Bethune *et al.* (4.77). A Raman spectrum recorded from the SWNT-containing material is shown in Fig. 4.27(e), where it is compared with spectra from various other carbon materials. The spectrum contains two sharp first order lines at 1566 and 1592 cm^{-1}. Sharp lines at these positions are not observed in the spectrum from arc-evaporated soot produced without cobalt,

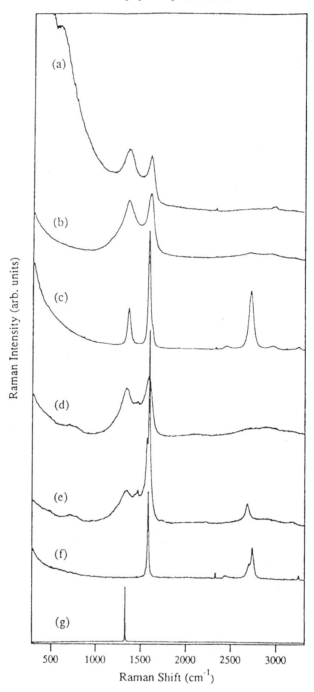

Fig. 4.27. Room temperature Raman spectra for a variety of carbon samples, from the work of Holden *et al.* (4.76). (a) Glassy carbon, (b) 'nanosoot' produced by laser pyrolysis, (c) nanosoot after heat treatment at 2820 °C, (d) soot produced by arc-evaporation, (e) soot produced by arc-evaporation with Co added to anode, (f) HOPG, (g) single-crystal diamond.

shown in Fig. 4.27(d), or in any of the other spectra from other carbons shown in Fig. 4.27. The authors showed that the lines at 1566 and 1592 cm^{-1} were revealed clearly in a 'difference spectrum' calculated by subtracting the spectrum of normal arc-evaporated soot from that of the cobalt-catalysed soot.

The observation of sharp features in this region is consistent with the predictions discussed above. As can be seen in Fig. 4.25, which refers to zig-zag tubes, theory predicts a line at $\sim$ 1590 cm^{-1}, which splits into a doublet in the low-diameter region relevant to single-walled nanotubes. Theory also predicts lines in this region for armchair and chiral nanotubes.

A Raman study of nanotube 'rope' samples was reported in early 1997 (4.78). A very interesting feature of this work was that the spectra were shown to vary according to the laser excitation frequency. These results were interpreted in terms of a resonant Raman scattering process, and provide evidence that tubes with different diameters couple with different efficiencies to the laser field. Resonant Raman scattering occurs when the energy of an incident photon matches the energy of strong optical absorption electronic transitions. This greatly increases the intensity of the observed Raman effect. Spectra recorded at five different laser frequencies are shown in Fig. 4.28. Dramatic changes in the distribution of the line frequencies can be seen, particularly in the case of the line at around 186 cm^{-1}, which is attributed to a radial breathing mode, in which all of the atoms in the tube are displaced inward or outward by an equal amount. These observations apparently show that the electronic density of states is dependant on nanotube diameter, and therefore provide further evidence for quantum behaviour in single-walled carbon nanotubes.

4.7 Electron energy loss spectroscopy of nanotubes

Electron energy loss spectroscopy (EELS) is a valuable technique for light element analysis in the transmission electron microscope, and can provide useful information on oxidation states and bonding (4.80). For graphite-related carbon specimens, information can be obtained from two energy ranges. In the approximate range 280–300 eV, energy loss is associated with the excitation of an electron from the 1s carbon core to the Fermi level. The nature of the 1s core level spectra can enable estimates to be made of the relative amounts of sp^2 and sp^3 bonding. In the region 0–40 eV, energy loss is due to the excitation of plasmons, arising from the π electrons and the π and σ states of the carbon L shell. Highly ordered graphite gives a broad peak in this region, centred on about 28 eV.

A number of EELS studies have been reported of both multiwalled and single-walled carbon nanotubes (e.g. 4.81–4.83). Ajayan, Iijima and Ichihashi

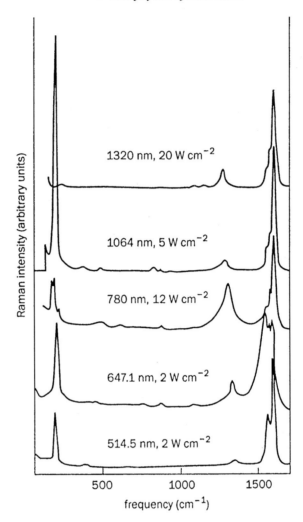

Fig. 4.28. Room temperature Raman spectra for single-walled nanotube 'rope' samples excited at five different laser frequencies (4.78, 4.79).

of NEC recorded EELS spectra from multilayer nanotubes of different diameters (4.81). They found that the bulk plasmon loss peak ($\sigma + \pi$) shifted to lower energy with decreasing diameter. This was taken to indicate that the π electrons become less delocalised as the tubes become smaller, due to curvature-induced strain, and therefore contribute less to the collective excitations. In the range 260–320 eV, in which the K shell excitations occur, the spectra from the tubes were broadly similar to those of graphite. For single-walled tubes, the bulk plasmon ($\sigma + \pi$) loss peak generally shifts to lower energies than for multiwalled tubes, although the precise position of this peak seems to be very sensitive to diameter (4.83).

4.8 Nanotube field emitters

There is currently considerable interest in the development of field-emitting cathodes, for a variety of potential microelectronic and display applications. In particular, much effort is being devoted to the fabrication of arrays of field emitters for use in flat screen displays. Most of the research in this area has focused on the deposition of metal cones onto silicon substrates using photolithographic techniques, although a variety of other possible fabrication methods have been proposed (4.84). The high conductivity, sharp tips and long, narrow shape of carbon nanotubes suggested to a number of groups that they might make useful field emitters, and some quite promising results have been achieved (4.85).

Some of the first experiments on field emission from nanotubes were carried out by Walt de Heer and colleagues (4.86). These workers used aligned nanotube films, prepared in the way described in Section 2.9. The arrangement they employed is shown in Fig. 4.29. The film of aligned nanotubes, roughly 2 mm in diameter, was supported on a Teflon substrate, and a 3 mm copper electron microscopy grid was held above it, at a distance of approximately 20 µm. The anode was situated about 1 cm above the copper grid, and current densities of up to 0.1 A cm^{-1} were reported, at a voltage of approximately 700 V. This level of emission should be sufficient to produce an image on a phosphor-coated display. Other work in this area includes a demonstration of field emission from a flat panel display employing a nanotube–epoxy composite as the source (4.87). There is also interest in fabricating aligned arrays of tubes using catalytic methods, as discussed in Chapter 2 (p. 34). This approach has produced some very promising results (4.88).

In addition to the work on arrays of nanotubes, field emission experiments have also been carried out on individual tubes. The earliest work in this area arose out of studies by Smalley's group into the use of single nanotubes as probes in scanning tunnelling or atomic force microscopy (see Chapter 6). During these studies they developed a method for mounting nanotubes on the tips of conventional carbon fibres, and this enabled them to determine the field emission characteristics of individual tubes (4.89). Interestingly, they found that field emission was dramatically enhanced when the tubes were opened by laser vaporisation or oxidation. The most intense emission appeared to originate from individual carbon chains which became unravelled from the edges of the opened tubes. Field emission from individual MWNTs has also been studied by Yahachi Saito and colleagues, who have used closed and open tubes as the sources in a field emission microscope (4.90). The motivation behind

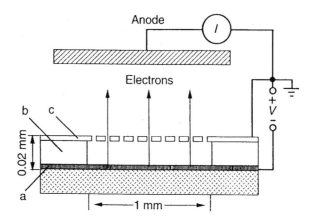

Fig. 4.29. Schematic illustration of the method used by de Heer *et al.* to measure the field emission properties of aligned nanotubes (4.86).

these studies is to explore the possibility of using individual nanotube field emitters in cathode ray tubes or as guns in electron microscopes.

4.9 Discussion

Much of this chapter has been concerned with the electronic properties of carbon nanotubes. This has been perhaps the most active area of nanotube research to date, and one in which some of the most spectacular advances have been made. Particularly notable are the conductivity experiments on individual nanotubes, which must represent some of the most impressive achievements in the emerging field of nanotechnology. The first of these measurements were made on multiwalled nanotubes, and showed that the resistivity could vary quite considerably from tube to tube. This appears to confirm the theoretical prediction that the electronic properties of nanotubes depend upon structure. Subsequently, measurements on single-walled nanotubes using a combination of STM and STS have provided direct proof of a link between electronic characteristics and structure. Experiments with individual tubes, both single-walled and multiwalled, have also provided evidence for quantum wire behaviour, as have other techniques such as Raman spectroscopy, and the construction of nanoscopic devices based on nanotubes has been reported.

These remarkable results have led to some speculation about nanotubes forming the basis of a future 'nanoelectronics'. It is certainly true that new approaches will be needed if miniaturisation is to continue at the rate we have witnessed over the past four decades. At present, the smallest structural features which are found on commercial silicon chips have dimensions of around 0.35 μm. Further miniaturisation to about 0.1 μm may be possible, but beyond that conventional photolithography hits a barrier. At some stage, the 'top-down' approach to device fabrication may have to be replaced by 'bottom-up'. There are many possible ways in which this might be achieved, all based in some way on carbon rather than silicon (4.91). Carbon nanotubes certainly have important advantages over some of the other systems which have been studied. They are chemically unreactive, robust against a Peierls distortion, and physically strong and resilient. However, the construction of a useful device would require enormous numbers of nanotubes to be positioned on a substrate with near-atomic accuracy. Is it possible to conceive of a way in which this might be achieved? Recent work by Hongjie Dai and colleagues at Stanford might provide a pointer (4.92). This group used silicon wafers patterned with micrometre-scale islands of an iron compound to catalyse the formation of single-walled nanotubes. The resulting tubes grew out from the catalytic regions, in some cases forming bridges between adjacent islands. Refinement of this technique might enable complex patterns of nanotubes to be formed on a substrate, although it hardly needs saying that the desired degree of control is still well beyond our reach.

While the electronic properties of carbon nanotubes have been probed in detail, other physical properties have received less attention. There have been relatively few studies of the magnetic properties of nanotubes, for example, and the results which have been obtained have not always agreed with theoretical predictions. Interesting optical properties have also been predicted for nanotubes, but again there has been little experimental work in this area. Overall there is still much to be learnt about the physics of carbon nanotubes.

References

(4.1) P. R. Wallace, 'The band theory of graphite', *Phys. Rev.*, **71**, 622 (1947).

(4.2) I. L. Spain, 'Electronic transport properties of graphite, carbons and related materials', *Chem. Phys. Carbon*, **16**, 119 (1981).

(4.3) J. C. Slonczewski and P. R. Weiss, 'Band structure of graphite', *Phys. Rev.*, **109**, 272 (1958).

(4.4) J. W. McClure, 'Band structure of graphite and de Haas–van Alphen effect', *Phys. Rev.*, **108**, 612 (1957).

(4.5) R. R. Haering and S. Mrozowski, 'Band structure and electronic properties of graphite crystals', *Prog. Semicond.*, **5**, 273 (1960).

(4.6) I. L. Spain, 'The electronic properties of graphite', *Chem. Phys. Carbon*, **8**, 1 (1973).

(4.7) B. T. Kelly, *The physics of graphite*, Applied Science Publishers, London and New Jersey, 1981.

(4.8) M.-F. Charlier and A. Charlier, 'The electronic structure of graphite and its basic origins', *Chem. Phys. Carbon*, **20**, 59 (1987).

(4.9)   M. S. Dresselhaus and G. Dresselhaus, 'Intercalation compounds of graphite', *Adv. Phys.*, **30**, 139 (1981).

(4.10) M. S. Dresselhaus, G. Dresselhaus, K. Sugihara, I. L. Spain and H. A. Goldberg, *Graphite fibers and filaments*, Springer-Verlag, Berlin, 1988.

(4.11) M. Endo, Y. Hishiyama and T. Koyama, 'Magnetoresistance effect in graphitising carbon fibres prepared by benzene decomposition', *J. Phys. D*, **15**, 353 (1982).

(4.12) P. Delhaes, 'Positive and negative magnetoresistances in carbons', *Chem. Phys. Carbon*, **7**, 193 (1971).

(4.13) R. Saito, M. Fujita, G. Dresselhaus and M. S. Dresselhaus, 'Electronic structure of graphene tubules based on C_{60}', *Phys. Rev. B*, **46**, 1804 (1992).

(4.14) M. S. Dresselhaus, G. Dresselhaus and R. Saito, 'Physics of carbon nanotubes', *Carbon*, **33**, 883 (1995).

(4.15) M. S. Dresselhaus, G. Dresselhaus and P. C. Eklund, *Science of fullerenes and carbon nanotubes*, Academic Press, San Diego, 1996.

(4.16) N. Hamada, S. Sawada and A. Oshiyama, 'New one-dimensional conductors: graphitic microtubules', *Phys. Rev. Lett.*, **68**, 1579 (1992).

(4.17) J. W. Mintmire, B. I. Dunlap and C. T. White, 'Are fullerene tubules metallic?', *Phys. Rev. Lett.*, **68**, 631 (1992).

(4.18) C. Kittel, *Solid state physics* 5th Edition, John Wiley, New York, 1976, p. 133.

(4.19) R. Saito, M. Fujita, G. Dresselhaus and M. S. Dresselhaus, 'Electronic structure of chiral graphene tubules', *Appl. Phys. Lett.*, **60**, 2204 (1992).

(4.20) R. E. Peierls, *Quantum theory of solids*, Oxford University Press, Oxford, 1955, p. 108.

(4.21) K. Harigaya and M. Fujita, 'Dimerization structures of metallic and semiconducting fullerene tubules', *Phys. Rev. B*, **47**, 16,563 (1993).

(4.22) R. Saito, G. Dresselhaus and M. S. Dresselhaus, 'Electronic-structure of double-layer graphene tubules', *J. Appl. Phys.*, **73**, 494 (1993).

(4.23) J.-C. Charlier and J.-P. Michenaud, 'Energetics of multilayered carbon tubules', *Phys. Rev. Lett.*, **70**, 1858 (1993).

(4.24) P. Lambin, L. Philippe, J.-C. Charlier and J.-P. Michenaud, 'Electronic band structure of multilayered carbon tubules', *Comp. Mat. Sci.*, **2**, 350 (1994).

(4.25) M. Heiblum and L. F. Eastman, 'Ballistic electrons in semiconductors', *Sci. Amer.*, **256(2)**, 65 (1987).

(4.26) M. F. Lin and K. W.-K. Shung, 'Magnetoconductance of carbon nanotubes', *Phys. Rev. B*, **51**, 7592 (1995).

(4.27) L. Chico, L. X. Benedict, S. G. Louie and M. L. Cohen, 'Quantum conductance of carbon nanotubes with defects', *Phys. Rev. B*, **54**, 2600 (1996).

(4.28) P. Lambin, J. P. Vigneron, A. Fonseca, J. B. Nagy and A. A. Lucas, 'Atomic structure and electronic properties of bent carbon nanotubes', *Synth. Met.*, **77**, 249 (1996).

(4.29) L. Chico, V. H. Crespi, L. X. Benedict, S. G. Louie and M. L. Cohen, 'Pure carbon nanoscale devices: Nanotube heterojunctions', *Phys. Rev. Lett.*, **76**, 971 (1996).

(4.30) P. G. Collins, A. Zettl, H. Bando, A. Thess and R. E. Smalley, 'Nanotube nanodevice', *Science*, **278**, 100 (1997).

(4.31) H. Ajiki and T. Ando, 'Electronic states of carbon nanotubes', *J. Phys. Soc. Japan*, **62**, 1255 (1993).

(4.32) H. Ajiki and T. Ando, 'Energy bands of carbon nanotubes in magnetic fields', *J. Phys. Soc. Japan*, **65**, 505 (1996).

(4.33) R. Saito, G. Dresselhaus and M. S. Dresselhaus, 'Magnetic energy bands of carbon nanotubes', *Phys. Rev. B*, **50**, 14 698 (1994). [Erratum: *Phys. Rev. B*, **53**, 10 408 (1996)].

(4.34) T. W. Ebbesen and P. M. Ajayan, 'Large-scale synthesis of carbon nanotubes' *Nature*, **358**, 220 (1992).

(4.35) L. Langer, V. Bayot, E. Grivei, J.-P. Issi, J. P. Heremans, C. H. Olk, L. Stockman, C. Van Haesendonck and Y. Bruynseraede, 'Quantum transport in a multiwalled carbon nanotube', *Phys. Rev. Lett.*, **76**, 479 (1996).

(4.36) H. Dai, E. W. Wong and C. M. Lieber, 'Probing electrical transport in nanomaterials: conductivity of individual carbon nanotubes', *Science*, **272**, 523 (1996).

(4.37) T. W. Ebbesen, H. J. Lezec, H. Hiura, J. W. Bennett, H. F. Ghaemi and T. Thio, 'Electrical conductivity of individual carbon nanotubes' *Nature*, **382**, 54 (1996).

(4.38) S. Frank, P. Poncharal, Z. L. Wang and W. A. de Heer, 'Carbon nanotube quantum resistors', *Science*, **280**, 1744 (1998).

(4.39) A. Thess, R. Lee, P. Nikolaev, H. Dai, P. Petit, J. Robert, C. Xu, Y. H. Lee, S. G. Kim, A. G. Rinzler, D. T. Colbert, G. E. Scuseria, D. Tománek, J. E. Fischer and R. E. Smalley, 'Crystalline ropes of metallic carbon nanotubes', *Science*, **273**, 483 (1996).

(4.40) J. E. Fischer, H. Dai, A. Thess, R. Lee, N. M. Hanjani, D. L. Dehaas and R. E. Smalley, 'Metallic resistivity in crystalline ropes of single-wall carbon nanotubes', *Phys. Rev. B*, **55**, R4921 (1997).

(4.41) M. Bockrath, D. H. Cobden, P. L. McEuen, N. G. Chopra, A. Zettl, A. Thess and R. E. Smalley, 'Single-electron transport in ropes of carbon nanotubes', *Science*, **275**, 1922 (1997).

(4.42) S. J. Tans, M. H. Devoret, H. Dai, A. Thess, R. E. Smalley, L. J. Geerligs and C. Dekker, 'Individual single-wall carbon nanotubes as quantum wires', *Nature*, **386**, 474 (1997).

(4.43) J. W. G. Wildöer, L. C. Venema, A. G. Rinzler, R. E. Smalley and C. Dekker, 'Electronic structure of atomically resolved carbon nanotubes', *Nature*, **391**, 59 (1998).

(4.44) T. W. Odom, J.-L. Huang, P. Kim and C. M. Lieber, 'Atomic structure and electronic properties of single-walled carbon nanotubes', *Nature*, **391**, 62 (1998).

(4.45) S. J. Tans, A. R. M. Verschueren and C. Dekker, 'Room- temperature transistor based on a single carbon nanotube', *Nature*, **393**, 49 (1998).

(4.46) R. S. Lee, H. J. Kim, J. E. Fischer, A. Thess and R. E. Smalley, 'Conductivity enhancement in single-walled carbon nanotube bundles doped with K and Br', *Nature*, **388**, 255 (1997).

(4.47) A. M. Rao, P. C. Eklund, S. Bandow, A. Thess and R. E. Smalley, 'Evidence for charge transfer in doped carbon nanotube bundles from Raman scattering' *Nature*, **388**, 257 (1997).

(4.48) R. C. Haddon, A. F. Hebard, M. J. Rosseinsky, D. W. Murphy, S. J. Duclos, K. B. Lyons, B. Miller, J. M. Rosamilia, R. M. Fleming, A. R. Kortan, S. H. Glarum, A. V. Makhija, A. J. Muller, R. H. Eick, S. M. Zaharak, R. Tycko, G. Dabbagh and F. A. Thiel, 'Conducting films of C_{60} and C_{70} by alkali metal doping', *Nature*, **350**, 320 (1991).

(4.49) G. Wagoner, 'Spin resonance of charge carriers in graphite', *Phys. Rev.*, **118**, 647 (1960).

(4.50) F. J. Dyson, 'Electron spin resonance absorption in metals. II. Theory of electron diffusion and the skin effect', *Phys. Rev.*, **98**, 349 (1955).

(4.51) I. C. Lewis and L. S. Singer, 'Electron spin resonance and the mechanism of carbonization', *Chem. Phys. Carbon*, **17**, 1 (1981).

(4.52) J. B. Jones and L. S. Singer, 'Electron spin resonance and the structure of carbon fibers', *Carbon*, **20**, 379 (1982)

(4.53) M. Kosaka, T. W. Ebbesen, H. Hiura and K. Tanigaki, 'Electron spin resonance of carbon nanotubes', *Chem. Phys. Lett.*, **225**, 161 (1994).

(4.54) M. Kosaka, T. W. Ebbesen, H. Hiura and K. Tanigaki, 'Annealing effect on carbon nanotubes. An ESR study', *Chem. Phys. Lett.*, **233**, 47 (1995).

(4.55) O. Chauvet, L. Forro, W. Bacsa, D. Ugarte, B. Doudin and W. A. de Heer, 'Magnetic anisotropies of aligned carbon nanotubes', *Phys. Rev. B*, **52**, R6963 (1995).

(4.56) P. Petit, E. Jouguelet, J. E. Fischer, A. G. Rinzler and R. E. Smalley, 'Electron spin resonance and microwave resistivity of single-wall carbon nanotubes', *Phys. Rev. B*, **56**, 9275 (1997).

(4.57) R. C. Haddon, 'Magnetism of the carbon allotropes', *Nature*, **378**, 249 (1995).

(4.58) H. Ajiki and T. Ando, 'Magnetic properties of carbon nanotubes', *J. Phys. Soc. Japan*, **62**, 2470 (1993). [Erratum, *J. Phys. Soc. Japan*, **63**, 4267 (1994).]

(4.59) H. Ajiki and T. Ando, 'Magnetic properties of ensembles of carbon nanotubes', *J. Phys. Soc. Japan*, **64**, 4382 (1995).

(4.60) M. F. Lin and W.-K. Shung, 'Magnetization of graphene tubules', *Phys. Rev. B*, **52**, 8423 (1995).

(4.61) A. P. Ramirez, R. C. Haddon, O. Zhou, R. M. Fleming, J. Zhang, S. M. McClure and R. E. Smalley, 'Magnetic susceptibility of molecular carbon: Nanotubes and fullerite', *Science*, **265**, 84 (1994).

(4.62) X. K. Wang, R. P. H. Chang, A. Patashinski and J. B. Ketterson, 'Magnetic-susceptibility of buckytubes', *J. Mater. Res.*, **9**, 1578 (1994).

(4.63) F. Huaxiang, Y. Ling and X. Xide, 'Optical properties for graphene microtubules of different geometries', *Solid State Commun.*, **91**, 191 (1994).

(4.64) X. G. Wan, J. M. Dong and D. Y. Xing, 'Symmetry effect on the optical properties of armchair and zigzag nanotubes', *Solid State Commun.*, **107**, 791 (1998).

(4.65) X. G. Wan, J. M. Dong and D. Y. Xing, 'Optical properties of carbon nanotubes', *Phys. Rev. B*, **58**, 6756 (1998).

(4.66) J. Jiang, J. M. Dong, X. G. Wan and D. Y. Xing, 'A new kind of nonlinear optical material: the fullerene tube', *J. Phys. B*, **31**, 3079 (1998).

(4.67) V. A. Margulis and T. A. Sizikova, 'Theoretical study of third-order nonlinear optical response of semiconductor carbon nanotubes', *Physica B*, **245**, 173 (1998).

(4.68) W. A. de Heer, W. S. Bacsa, A. Châtelain, T. Gerfin, R. Humphrey-Baker, L. Forro and D. Ugarte, 'Aligned carbon nanotube films: production and optical and electronic properties', *Science*, **268**, 845 (1995).

(4.69) R. A. Jishi, L. Venkataraman, M. S. Dresselhaus and G. Dresselhaus, 'Phonon modes in carbon nanotubules', *Chem. Phys. Lett.*, **209**, 77 (1993).

(4.70) R. A. Jishi, D. Inomata, K. Nakao, M. S. Dresselhaus and G. Dresselhaus, 'Electronic and lattice properties of carbon nanotubes', *J. Phys. Soc. Japan*, **63**, 2252 (1994).

(4.71) R. A. Jishi, L. Venkataraman, M. S. Dresselhaus and G. Dresselhaus, 'Symmetry properties of chiral carbon nanotubes', *Phys. Rev. B*, **51**, 11,176 (1995).

(4.72) P. C. Eklund, J. M. Holden and R. A. Jishi, 'Vibrational-modes of carbon nanotubes – spectroscopy and theory', *Carbon*, **33**, 959 (1995).

(4.73) H. Hiura, T. W. Ebbesen, K. Tanigaki and H. Takahashi, 'Raman studies of carbon nanotubes', *Chem. Phys. Lett.*, **202**, 509 (1993).

(4.74) J. Kastner, T. Pichler, H. Kuzmany, S. Curran, W. Blau, D. N. Weldon, M. Delamesiere, S. Draper and H. Zandbegen, 'Resonance Raman and infrared spectroscopy of carbon nanotubes', *Chem. Phys. Lett.*, **221**, 53 (1994).

(4.75) W. S. Bacsa, D. Ugarte, A. Châtelain and W. A. de Heer, 'High-resolution electron microscopy and inelastic light scattering of purified multishelled carbon nanotubes', *Phys. Rev. B*, **50**, 15,473 (1994).

(4.76) J. M. Holden, P. Zhou, X.-X. Bi, P. C. Eklund, S. Bandow, R. A. Jishi, K. D. Chowdhury, G. Dresselhaus and M. S. Dresselhaus, 'Raman-scattering from nanoscale carbons generated in a cobalt-catalyzed carbon plasma', *Chem. Phys. Lett.*, **220**, 186 (1994).

(4.77) D. S. Bethune, C. H. Kiang, M. S. de Vries, G. Gorman, R. Savoy, J. Vasquez and R. Beyers, 'Cobalt-catalysed growth of carbon nanotubes with single-atomic-layer walls', *Nature*, **363**, 605 (1993).

(4.78) A. M. Rao, E. Richter, S. Bandow, B. Chase, P. C. Eklund, K. A. Williams, S. Fang, K. R. Subbaswamy, M. Menon, A. Thess, R. E. Smalley, G. Dresselhaus and M. S. Dresselhaus 'Diameter-selective Raman scattering from vibrational modes in carbon nanotubes', *Science*, **275**, 187 (1997).

(4.79) M. S. Dresselhaus, G. Dresselhaus, P. C. Eklund and R. Saito, 'Carbon nanotubes', *Physics World*, **11(1)**, 33 (1998).

(4.80) R. F. Egerton, *Electron energy loss spectroscopy in the electron microscope* (2nd Ed.), Plenum, New York and London, 1996.

(4.81) P. M. Ajayan, S. Iijima and T. Ichihashi, 'Electron-energy-loss spectroscopy of carbon nanometer-size tubes', *Phys. Rev. B*, **47**, 6859 (1993).

(4.82) R. Kuzuo, M. Terauchi and M. Tanaka 'Electron energy-loss spectra of carbon nanotubes', *Jap. J. Appl. Phys.*, **31**, L1484 (1992).

(4.83) R. Kuzuo, M. Terauchi, M. Tanaka and Y. Saito, 'Electron energy-loss spectra of single-shell carbon nanotubes', *Jap. J. Appl. Phys.*, **33**, L1316 (1994).

(4.84) I. Brodie and C. A. Spindt, 'Vacuum microelectronics', *Adv. Electron. Electron Phys.*, **83**, 1 (1992).

(4.85) R. F. Service, 'Nanotubes show image-display talent', *Science*, **270**, 1119 (1995).

(4.86) W. A. de Heer, A. Châtelain and D. Ugarte, 'A carbon nanotube field-emission electron source', *Science*, **270**, 1179 (1995).

(4.87) Q. H. Wang, A. A. Setlur, J. M. Lauerhaas, J. Y. Dai, E. W. Seelig and R. P. H. Chang, 'A nanotube-based field-emission flat panel display', *Appl. Phys. Lett.*, **72**, 2912 (1998).

(4.88) Z. F. Ren, Z. P. Huang, J. W. Xu, J. H. Wang, P. Bush, M. P. Siegal and P. N. Provencio, 'Synthesis of large arrays of well-aligned carbon nanotubes on glass', *Science*, **282**, 1105 (1998).

(4.89) A. G. Rinzler, J. H. Hafner, P. Nikolaev, L. Lou, S. G. Kim, D. Tománek, P. Nordlander, D. T. Colbert and R. E. Smalley, 'Unravelling nanotubes: field emission from an atomic wire', *Science*, **269**, 1550 (1995).

(4.90) Y. Saito, K. Hamaguchi, K. Hata, K. Uchida, Y. Tasaka, F. Ikazaki, M. Yumura, A. Kasuya and Y. Nishina, 'Conical beams from open nanotubes', *Nature*, **389**, 554 (1997).

(4.91) D. Bradley, 'Crays the size of a paperback', *Chem. Soc. Rev.*, **24**, 379 (1995).

(4.92) J. Kong, H. T. Soh, A. M. Cassell, C. F. Quate and H. Dai, 'Synthesis of individual single-walled carbon nanotubes on patterned silicon wafers', *Nature*, **395**, 878 (1998).

5

Nanocapsules and nanotest-tubes

Daedalus . . . is now seeking ways of incorporating 'windows' into
their structure so that they can absorb or exchange internal
molecules . . .
David Jones, *New Scientist*, 3 November, 1966

When Daedalus dreamed up hollow graphitic molecules in the mid-sixties he speculated that they would make ideal molecular containers, and so it has proved. The first demonstration of this came just a few days after the discovery of C_{60} in September 1985 when Kroto, Smalley and their colleagues prepared the first metallofullerene, La@C_{60} (5.1). This not only provided further, perhaps definitive, proof that the cage hypothesis was correct, but also marked the beginning of a whole new field of molecular science. Subsequent work has shown that many different atoms ranging from rare gases to uranium can be encapsulated inside fullerenes (5.2, 5.3), although testing the properties of these materials has been difficult, since they can currently only be produced in small amounts. There is also interest in the idea of chemically creating openings in fullerenes, to allow foreign molecules in and out (5.4), in exactly the way envisaged by Daedalus.

When carbon nanotubes and nanoparticles arrived it was natural that attempts would be made to introduce foreign materials into the central cavities. Two approaches were tried, and both proved successful. The first involved carrying out arc-evaporation in the usual way, but with an anode containing some of the material to be encapsulated. This technique generally seems to favour the formation of filled nanoparticles rather than nanotubes, and is only applicable to materials which can survive the extreme conditions of the electric arc. The second approach was to open the cages by chemical means and then rely on capillarity to draw material inside. This has proved an excellent method of filling nanotubes, and can be applied to relatively fragile materials, including biological molecules. Research into the opening and filling of carbon nanotubes and nanoparticles has been one of the most active fields of fullerene science, and may lead to applications in areas as diverse as fine particle magnetism and biosensors. In this chapter the methods used to open and fill carbon nanotubes and nanoparticles, and their possible applications, will be discussed in detail. Firstly though, a brief review of endohedral metallofullerenes will be given.

5.1 Metallofullerenes

When considering which atom might be a good candidate for encapsulation, Kroto, Smalley and colleagues initially chose iron, believing that an iron atom might fit as comfortably inside a C_{60} molecule as it does in the 'sandwich compound' ferrocene. However this was unsuccessful, so they discussed possible alternatives. Smalley's student Jim Heath was convinced that lanthanum would be a good candidate, arguing that the inside of a buckyball might resemble the electron-rich environment which lanthanum occupies in LaF_6. This proved to be an inspired choice, and single lanthanum atoms were successfully encapsulated in C_{60} just a week after the initial observation of C_{60} itself (5.1). The method used was to vaporise lanthanum-impregnated graphite under the same conditions used to produce C_{60}. A short time later, atoms of calcium, barium and strontium were also encapsulated. It was clear that a new branch of organometallic chemistry had been born, with the potential for producing a whole range of new molecular materials with a wide variety of properties, but significant further progress had to await the discovery of bulk synthesis methods.

The first macroscopic amounts of endohedral metallofullerenes were prepared in 1991, not using the Krätschmer–Huffman technique, but using an apparatus developed by Smalley's group to vaporise graphite in a helium-filled tube oven at high temperature. When pure graphite rods were replaced with composite targets made from powdered graphite and lanthanum oxide, lanthanum-containing fullerenes were produced in milligram quantities (5.5). As well as $La@C_{60}$ (where the @ symbol denotes encapsulation), these included $La@C_{70}$, $La@C_{74}$ and $La@C_{82}$, the most stable of which turned out to be $La@C_{82}$. The reasons for the rather surprising stability of $La@C_{82}$ are not fully understood, but probably relate to the electronic structures of the fullerene molecules. Hückel calculations show that C_{60} has a closed electronic shell, so that electrons donated from an encapsulated metal atom will go into the antibonding orbitals, making the molecule less stable (5.6). On the other hand, Hückel calculations indicate that C_{82} is two electrons short of a completely filled configuration, so that the donation of two electrons from a metal atom would produce a stable, closed shell structure. Unfortunately for this simple model, the La atom in La@C_{82} is believed to have a valency of 3, which would leave one unpaired electron (5.7). There is clearly a gap in our understanding of why certain fullerenes make better 'containers' than others. Progress in this area might help in understanding the filling of nanotubes and other fullerene-related carbons, which is discussed below. There is now a substantial literature on metallofullerenes, and a number of reviews have been published (5.2, 5.3).

5.2 Filling nanotubes and nanoparticles by arc-evaporation

5.2.1 Early work

The first attempts to put foreign material inside nanotubes used the arc-evaporation technique to vaporise a mixture of graphite and lanthanum oxide, taking a lead from Heath's work on C_{60}. Thus, instead of using pure graphite electrodes, the anode was drilled out and a mixture of lanthanum oxide and graphite powder inserted. Arc discharge was carried out in the usual way, and the carbon deposited on the cathode collected. Experiments of this kind were carried out simultaneously by Rodney Ruoff's group in the US (5.8) and Yahachi Saito's group in Japan (5.9) and produced identical results: instead of filled nanotubes, the cathodic soot contained significant numbers of filled nanoparticles, as shown in Fig. 5.1 (quite why nanoparticles were favoured over nanotubes in this way remains unclear). A micrograph of an individual filled nanoparticle is shown in Fig. 5.2. Lattice imaging and electron diffraction showed that the material inside the nanoparticles was single-crystal lanthanum carbide, LaC_2, rather than pure lanthanum. A striking aspect of this work was that the usually hygroscopic LaC_2 was apparently protected against reaction with moisture even after several days exposure to air, demonstrating that the nanocapsules were completely sealed.

5.2.2 Further studies

Since the early work, many studies of filled carbon nanocages prepared by the modified electrode technique have been carried out (5.10–5.24). In most of these studies, filled nanoparticles were more commonly seen than filled nanotubes, although the latter were also found. A beautiful example of a filled nanotube prepared by arc-evaporation, from the work of Youichi Murakami and colleagues (5.13), is shown in Fig. 5.3. It has been shown that not all metals are capable of being encapsulated, and that different elements can have a

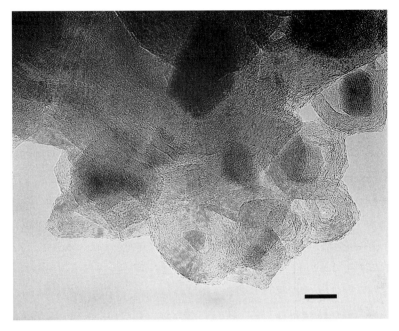

Fig. 5.1. Lanthanum carbide crystals encapsulated inside carbon nanoparticles, prepared using arc-evaporation (5.10). Scale bar 10 nm.

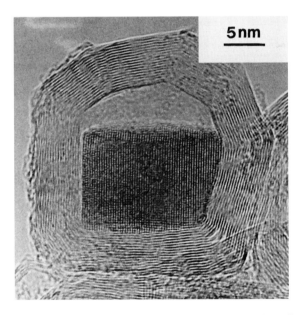

Fig. 5.2. Individual nanoparticle filled with lanthanum carbide, from the work of Yahachi Saito.

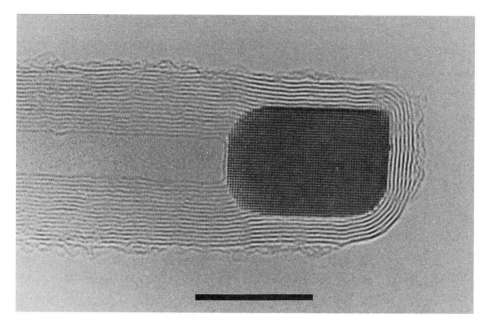

Fig. 5.3. Nanotube encapsulating crystal of tantalum carbide (5.13). Scale bar 10 nm.

profoundly different effect on the type of material produced by arc-evaporation. The most detailed studies in this area have been carried out by Saito's
group (5.10, 5.11, 5.14–5.17) and by Supapan Seraphin and colleagues at the
University of Arizona (5.12, 5.18, 5.19). Between them, these groups have
carried out encapsulation experiments with most of the metallic elements in
the periodic table (plus boron and silicon). A substantial programme of work,
concentrating mainly on the encapsulation of magnetic materials, has also
been carried out by Sara Majetich and colleagues of Carnegie Mellon University (5.20–5.22). A micrograph of encapsulated Co particles prepared by this
group is shown in Fig. 5.4. Their work is discussed further below.

   Seraphin and co-workers found that the elements they studied could be
grouped into four categories (5.12):

1 Elements which are encapsulated in the form of their carbides (B, V, Cr, Mn, Y, Zr,
   Nb, Mo).
2 Elements which are not encapsulated, but which tolerate the formation of carbon
   nanoparticles and nanotubes (Cu, Zn, Pd, Ag, Pt).
3 Elements which form stable carbides, consuming the carbon supply for nanostructure formation (Al, Si, Ti, W).
4 The iron-group metals Fe, Co, Ni, which stimulate the formation of single-walled
   nanotubes and 'nanobeads' in conventional arc-evaporation.

The encapsulated particles were usually found in the cathodic soot, or 'slag', although in the case of boron, coated particles were found in the soot deposited on the walls of the chamber. As Seraphin and colleagues point out, this may be related to the fact that boron is known to act as a graphitization catalyst. In the case of the iron-group metals Fe, Co, Ni, Seraphin and colleagues achieved encapsulation using specialised techniques which had previously been developed by Dravid *et al.* (5.23). These involved employing a vertical electrode geometry in which the anode consisted of a graphite crucible filled with the metal, and directing a jet of helium at the gap between the two electrodes during arcing. The filled nanoparticles were found in the soot deposited on the walls of the evaporation vessel. X-ray diffraction showed that the filling was the metal phase rather than carbide.

Saito and co-workers carried out arc-evaporation experiments with a range of rare-earth metals, and found that most of these (Sc, Y, La, Ce, Pr, Nd, Gd, Tb, Dy, Ho, Er, Tm and Lu) were encapsulated as carbides, although some (Sm, Eu and Yb) were not (5.9–5.11, 5.15, 5.16). They also succeeded in encapsulating Fe, Co and Ni. For these metals, the encapsulated particles were found both in the cathodic soot and in the wall soot, although in the latter case the nanocrystals were enclosed in amorphous carbon rather than well-formed graphite, as in Fig. 5.4. Unlike Seraphin and colleagues, Saito's group did not employ specialised arc-evaporation techniques to achieve the encapsulation of the iron-group metals, but in some cases the encapsulated material was carbide rather than pure metal. Encapsulation of platinum-group metals was also reported by this group (5.17).

Although arc-evaporation with modified electrodes usually seems to favour the formation of filled nanoparticles rather than nanotubes, a French group have studied the preparation of filled nanotubes in this way (5.24). They have successfully used arc-evaporation to produce nanotubes filled with a range of transition elements and rare earth elements, and in many cases find that almost complete filling of the tubes can be achieved.

The factors which affect whether or not particular elements are capable of being encapsulated have been discussed by a number of groups (5.10–5.12, 5.22, 5.24). Saito has suggested that the vapour pressure of the encapsulated elements may be important, pointing out that the rare-earth metals capable of being encapsulated belong to the group of relatively low vapour pressure metals, while those not encapsulated generally had a higher vapour pressure. Certainly, one might expect refractory materials to be more readily encapsulated in the cathodic soot than more volatile ones. However, as Seraphin and colleagues point out, this criterion does not appear to hold for the transition metals (5.12). Metals such as V, Cr, Mn and Y, which have high vapour pressures, have all been successfully encapsulated. Majetich and colleagues

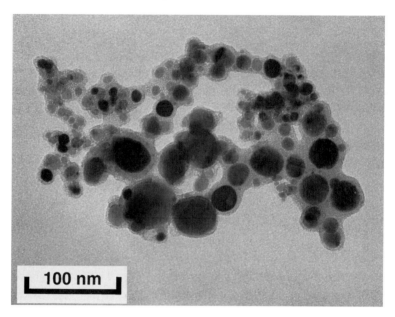

Fig. 5.4. Encapsulated cobalt nanocrystals found in the chamber soot following arc-evaporation. Courtesy Sara Majetich.

(5.20) have suggested that encapsulation will not occur if the materials have low melting points, or if their carbides have high melting points, but again this does not appear to apply in all cases. Seraphin and colleagues have suggested that the tendency of an element to form a carbide might be crucial in determining whether encapsulation occurs (5.12), although exceptions also exist to this rule. The French group proposed that the electronic structure of the metals may be important in determining whether or not they can be encapsulated (5.24). Clearly, no consensus exists at present on these issues, and further work is needed.

5.3 Preparation of filled nanoparticles from microporous carbon

The present author and colleagues have recently developed a simple method of preparing filled carbon nanoparticles which does not rely on arc-evaporation (5.25, 5.26). This technique arose out of studies of non-graphitizing carbon (see p. 28), which showed that high temperature heat treatments of these materials can result in the formation of closed carbon nanoparticles. In order to prepare filled carbon nanoparticles, the non-graphitizing carbon was firstly impregnated with a salt of the metal to be encapsulated. For example, to produce nanoparticles filled with molybdenum, the carbon was

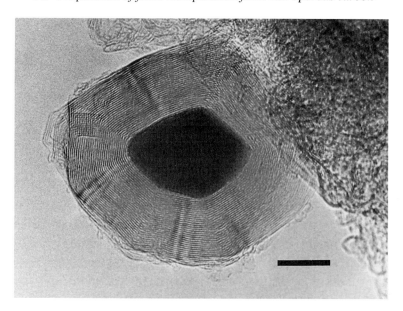

Fig. 5.5. Carbon nanoparticle filled with molybdenum carbide, prepared by heating an impregnated non-graphitizing carbon (5.26). Scale bar 10 nm.

impregnated with ammonium heptamolybdate. The dried product was then heated at temperatures in the range 1800–2500 °C, to produce a material containing many filled nanoparticles. It should be noted that the filled carbon nanoparticles can be produced at temperatures much lower than those needed to form nanoparticles from the pure carbon. This indicates that the particles of encapsulated material act as templates which promote formation of the graphite layers.

A typical filled carbon nanoparticle prepared by heating an impregnated non-graphitizing carbon is shown in Fig. 5.5. In this case the filling is molybdenum carbide. The particles produced in this way are very similar to those prepared by arc-evaporation. They range in size from about 10 nm to 500 nm. In some cases gaps are observed between the enclosed material and the carbon shell, while in other cases the cages were completely filled. At present, the yield of filled carbon nanoparticles produced by this method is rather low, and optimisation of the technique is needed. If this can be achieved, this approach may prove to be a much more practical route to filled nanoparticles than arc-evaporation.

5.4 Properties of filled nanoparticles

5.4.1 Protection from environmental degradation

The capability of carbon nanoparticles to protect encapsulated hygroscopic LaC_2 against reaction with moisture was noted in the early studies (see above). Subsequent work has confirmed the remarkable ability of nanoparticles to protect encapsulated material against environmental attack. Thus, Yahachi Saito has shown that hygroscopic carbides encapsulated in nanoparticles remain stable even after a year's exposure to air (5.11). Saito's group also carried out an experiment designed specifically to test the protective nature of the carbon shells (5.16), which involved exposing encapsulated iron particles to an environment with a relative humidity of 85% at a temperature of 80 °C for 7 days. After this period, X-ray diffraction showed no evidence for oxidation of the iron, while a control sample, containing naked iron particles with similar diameters, was substantially oxidised by the same treatment. Yositaka Yosida of Iwaki Meisei University has also shown that CeC_2 encapsulated in nanoparticles is stable in hot concentrated sulphuric acid, even though pristine CeC_2 would be highly unstable under these conditions (5.27). The chemical inertness of encapsulated nanoparticles could be very important in applications such as those discussed in the following two sections.

5.4.2 Encapsulation of magnetic materials

It has been suggested that carbon-coated magnetic nanoparticles might have important applications in areas such as magnetic data storage, xerography and magnetic resonance imaging (5.22). The role of the carbon layer would be to isolate the particles magnetically from each other, thus avoiding the problems caused by interactions between closely spaced magnetic bits, and to provide oxidation resistance. In addition the lubricating properties of the graphitic coatings might be helpful in magnetic recording applications. The potential of important applications has motivated a significant amount of research on the encapsulation of magnetic materials in carbon nanoparticles. Several groups have shown that the ferromagnetic metals Fe, Co and Ni retain their magnetic properties when encapsulated in carbon nanoparticles (5.19–5.23). In addition, Majetich and colleagues have demonstrated superparamagnetism in carbon-coated Co nanoparticles (5.21, 5.22). This group has also explored the use of arc-evaporation to encapsulate alloy magnets. For example, they have used a Sm_2Co_7 metallic precursor to produce Sm–Co–C particles. These alloys have large magnetic moments and anisotropy constants, and fine particles of such materials should therefore have more valuable properties than those of the ferromagnetic transition metals. Work in this important area will undoubtedly continue.

5.4.3 Encapsulation of radioactive materials

Another potential application for filled nanoparticles is in the area of radioactive materials. If radionuclides could be enclosed in seamless carbon nanocapsules, and the resulting nanoparticles were shown to be stable, this might provide a new method of nuclear waste disposal. Alternatively, radioisotopes encapsulated in carbon nanoparticles might be useful in nuclear medicine, either in radiotherapy or in imaging (see also next section). These ideas have been around for some time, but only a limited amount of work on this topic has been published in the open literature. The uranium metallofullerene $U@C_{28}$ was prepared by workers from Smalley's group in 1992 (5.28), and a number of endohedral metallofullerenes containing radioactive atoms have subsequently been prepared. For example, Japanese workers prepared radioactive ^{159}Gd and ^{161}Tb atoms in C_{82} by the irradiation of $Gd@C_{82}$ in a nuclear reactor (5.29). The metallofullerenes were found to be unaffected by the recoil energy from the β-decay of the encapsulated metal atoms. A short time later, another Japanese group prepared actinide carbides encapsulated in carbon nanoparticles (5.30). Thorium carbide was encapsulated by arc-evaporating graphite rods containing thoriated tungsten, while uranium carbide-containing nanoparticles were produced using rods filled with uranium ore. Nanoparticles filled with these radioactive materials were found to be stable for over a year. Encapsulation of uranium carbide in carbon nanoparticles using arc-evaporation methods has also been demonstrated by Argentinian and French workers (5.31). Again, the filled nanoparticles were found to be quite stable. It should be noted that nanotubes have been filled with uranium dioxide using chemical methods by Edman Tsang and colleagues at Oxford (5.32), as discussed below.

The effect of neutron irradiation on the structure of carbon nanoparticles containing molybdenum carbide was investigated by a Japanese group led by Atsuo Kasuya (5.33). The particles were irradiated in a nuclear reactor for 6 days at a flux of 10^{14} N cm^{-2} s^{-1}. It was shown by γ-ray spectrometry that the irradiated sample contained ^{99}Mo, which had been transformed from ^{98}Mo, and which decayed to ^{99}Tc. Electron microscopy showed that the carbon capsules suffered some damage following irradiation, although this appeared to be a result of nuclear reactions in the molybdenum carbide rather than a direct effect of the neutron irradiation.

There is undoubtedly growing interest in the encapsulation of radioactive materials in carbon nanocapsules. However, there are obvious safety problems with the use of the arc-evaporation method for the encapsulation of radionucl-ides, and alternative methods such as those involving the heat treatment of impregnated carbons might be more appropriate in this area.

5.5 Technegas

A discussion of filled carbon nanoparticles would be incomplete without some mention of the medical imaging agent 'Technegas'. The development of this product began in the late 1970s and early 1980s when a group at the Australian National University in Canberra were working on the preparation of a ventila-tion agent containing the radionuclide technetium-99m (where m stands for metastable). The aim was to produce a gas which could be inhaled and then used to image the lungs using a camera sensitive to γ-rays. In 1984 they found that this could be achieved using a technique which involved the simultaneous vaporisation of ^{99}Tc-m and graphite into an argon atmosphere (5.34). The aerosol produced in this way was sufficiently stable for it to be inhaled by a patient, and the technique proved to be of great value in diagnosing lung diseases; today, Technegas-based imaging is used throughout the world.

Despite its success as a diagnostic agent, the precise nature of the ^{99}Tc-m-carrying particles was at first unclear. In their original paper (5.34), the Canberra group suggested that Technegas was a radioactive soot, but did not propose a detailed structure. When C_{60} was discovered in 1985, Bill Burch, the group leader, was struck by the similarity of the method used to make fullerenes to the Technegas production technique. It began to look very likely that Technegas might be fullerene-related. (This raised the possibility that the Technegas patents might actually cover fullerene production (5.35).) Initial speculation centred on the idea that the aerosol consisted of ^{99}Tc-m-contain-ing metallofullerenes. However, more recent work using high resolution elec-tron microscopy (e.g. 5.36) has shown that the ^{99}Tc-m particles in Technegas are in fact relatively large crystals, typically 10 nm to 100 nm in diameter, coated with carbon that can be either graphitic or amorphous. Thus, the use of Technegas in medicine probably represents the first commercial application of filled carbon nanoparticles.

5.6 Opening and filling of nanotubes using chemical methods and capillarity

5.6.1 The work of Ajayan and Iijima

While Saito, Ruoff and their co-workers were attempting to make filled nanotubes using modified anodes, Pulickel Ajayan and Sumio Iijima tried a different line of attack, with striking success (5.37). Theoretical work had suggested that opened nanotubes should act as 'nanopipettes', sucking liquid inside by capillary action (5.38). Ajayan and Iijima decided to test this idea by treating a sample of tubes with molten lead, hoping that some of the lead might be drawn inside open tubes. Their method involved depositing particles of lead onto the tubes in a vacuum using electron-beam evaporation and then heating in air at 400 °C, a temperature sufficient to melt the lead. When they examined the resulting samples using TEM they found that a small proportion of the nanotubes had clearly been filled; examples are shown in Fig. 5.6. The fillings extended for distances up to a few hundred nanometres, but were frequently blocked by the presence of internal caps. Energy dispersive X-ray analysis of the filled tubes confirmed the presence of lead, but the exact nature of the material inside the tubes was difficult to establish. In general the filling appeared disordered, but where lattice images could be obtained they did not correspond with any known lead compound, raising the possibility that the structure might be affected by its extreme confinement.

In Ajayan and Iijima's work the proportion of nanotubes filled with lead was around 1%. The proportion of open tubes in a fresh sample is only around one in a million, so it is clear that the lead was not entering the tubes through existing holes. It seems that the tubes were opened by a chemical reaction involving the lead, oxygen and the carbon making up the tubes, but the exact nature of this reaction is unclear. One possibility is that the lead acted as a catalyst for the oxidation of carbon by oxygen. Subsequent work showed that molten material could be introduced into nanotubes which had previously been opened by oxidation, as discussed in Section 5.6.5.

5.6.2 *Selective opening using gas-phase oxidants*

The chief importance of Ajayan and Iijima's beautiful work was to demonstrate that oxidation (possibly catalysed by a metal) could be used to selectively remove the tips of nanotubes, while leaving the tube walls unaffected. However, like the modified-electrode method, it was not a universally applicable technique. It seemed that some way was needed to selectively open the tubes without simultaneously drawing material inside, leaving the tubes open for the subsequent introduction of foreign materials. The first part of this turned out to be remarkably simple. Work carried out simultaneously by Iijima's group in Japan and by Edman Tsang, Malcolm Green and the present author in Oxford showed that nanotubes could be opened with a reasonable degree of selectivity simply by heating in a mildly oxidising environment (5.39, 5.40). The Oxford group used carbon dioxide as the oxidising agent, making use of the 'reverse Boudouard reaction':

$$C_{(s)} + CO_{2(g)} \rightarrow 2CO_{(g)}.$$

Fig. 5.6. Nanotubes filled with lead or lead oxide, prepared using capillarity by Ajayan and Iijima (5.37).

Samples of nanotubes were heated at a range of temperatures up to 950 °C in CO_2 for periods up to 24 hours. When the oxidised samples were examined using TEM, it was found that in a small but significant number of cases the tube caps had been selectively attacked, frequently being thinned at the extremity, and sometimes completely opened. A tube which has been attacked in the cap region, but not completely opened is shown in Fig. 5.7(a), while an example of a typical opened nanotube is shown in Fig. 5.7(b).

The Japanese group heated the tubes in oxygen at temperatures up to 850 °C and observed similar behaviour. At temperatures below about 700 °C, little oxidation was observed, but at higher temperatures oxidation occurred rapidly: at 850 °C the entire sample was consumed after 15 minutes. Samples containing opened tubes could be prepared by heating in oxygen at 700 °C for short periods (typically 10 minutes). Using oxygen rather than carbon dioxide as the oxidising agent resulted in a higher proportion of open nanotubes, but the treatment was considerably more destructive, with many tubes being massively corroded.

Having succeeded in preparing opened nanotubes, both groups then endeavoured to fill the tubes with inorganic materials, but this proved far more difficult than expected. The Oxford group attempted to introduce solutions of

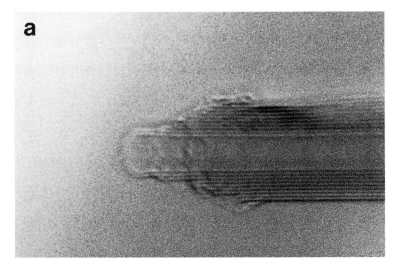

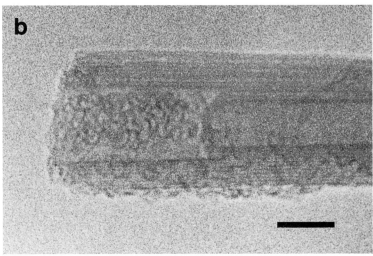

Fig. 5.7. Nanotubes following reaction with CO_2 (5.39). (a) Nanotube with outer layers stripped off in the tip region. (b) Completely opened tube with amorphous material inside. Scale bar 5 nm.

metal salts into the tubes using 'incipient wetness' techniques of the kind used in preparing supported catalysts, with the aim of drying off the solvent and reducing the salt to leave metal crystallites inside the tube. This met with little success, as did experiments by the Japanese group aimed at filling opened tubes with molten lead. It was not immediately clear why the two-step approach to filling nanotubes failed where the one-step method used by Ajayan and Iijima to fill tubes with molten lead succeeded so brilliantly. One possibility is that tubes become blocked by amorphous carbon as soon as they are opened, making subsequent filling difficult. This can be seen in Fig. 5.7(b), where an opened tube has been filled with debris up to an internal cap. However, the tendency of a particular liquid to enter an opened nanotube will also depend on its surface tension, as discussed in subsequent sections.

5.6.3 Opening by treatment with nitric acid

Following the difficulties with the carbon dioxide oxidation method, Tsang, Green and their colleagues in Oxford looked for alternative ways to open nanotubes. Previous work on the treatment of other fullerene-related carbons with nitric acid had suggested that this might be an effective way of selectively attacking pentagonal rings, so a similar treatment was applied to nanotubes. This proved to be highly successful (5.32). Not only did the nitric acid treatment result in a high yield of opened tubes, but the selectivity with which the tube tips were attacked was extraordinary. This is illustrated in Fig. 5.8, which shows micrographs of tubes from a sample treated with boiling nitric acid for 4.5 hours. In Fig. 5.8(a) the tube cap has been attacked at two points, labelled X and Y, both of which are at positions where pentagonal rings would have been present. In Fig. 5.8(b) the tube cap has been opened, and the internal caps have been selectively removed, so that a passage exists to the central cavity, although the remainder of the tube remains intact. This indicates that the acid has reacted only with the pentagonal rings; even where the edges of graphite sheets have been exposed, these have not been attacked, and very little thinning or stripping of the outer layers is observed. A similar effect is shown in Fig. 5.8(c).

The question arises of why the nitric acid method is so exquisitely selective. Once the reactive edges of the basal planes have been exposed, why are they not rapidly consumed, as is the case with gaseous oxidants? The answer is probably that a reaction occurs between the exposed edges and the nitric acid, resulting in the formation of surface carboxylate [–(COOH)] and other groups which act as a barrier to further reaction (see also next section). A similar reaction probably also occurs between the nitric acid and the outer tube. These hydrophillic groups would be expected to facilitate the filling of tubes with aqueous solutions, and with biological material as discussed below.

The nitric acid method could also be used to introduce foreign materials into the tubes in a one-stage process. Thus, when nickel nitrate was added to the nitric acid used in the oxidation treatment, tubes containing crystalline nickel oxide resulted. Tsang and colleagues also showed that nanotubes opened using the nitric acid method, unlike those opened by the less selective gas-phase oxidation methods, did not appear to be blocked with amorphous material.

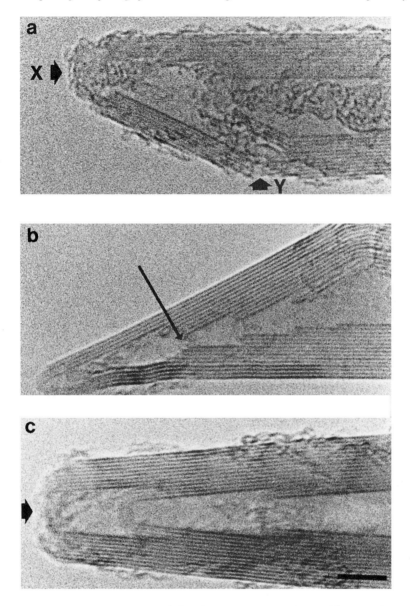

Fig. 5.8. Typical nanotube caps following treatment with boiling nitric acid. (a) Micrograph showing selective attack at points X and Y, where non-six-membered rings are present. (b) and (c) Micrographs showing destruction of multiple internal caps (5.32). Scale bar 5 nm.

Therefore, opened tubes could be subsequently filled with a whole range of materials simply by soaking the tubes in a solution or suspension of the substance, or of a precursor. This represented a significant advance.

.

5.6.4 Alternative liquid-phase oxidants

At about the same time as the work on nitric acid oxidation described above, other groups described tube opening using liquid-phase oxidants. Hidefumi Hiura, Thomas Ebbesen and Katsumi Tanigaki of the NEC corporation reported that oxidation with potassium permanganate in acid solution resulted in both the removal of nanoparticles and the opening of a significant proportion of the tubes (5.41). These workers also carried out XPS analysis of the oxidised tubes and found that approximately 15% of the surface had been converted to hydroxyl, carbonyl and carboxylic groups. A comprehensive study of the effect of various oxidants on nanotube samples was described by a Taiwanese chemist, Kuo Chu Hwang (5.42). Good results were achieved with $KMnO_4$, OsO_4 and RuO_4, but none of these reagents appeared to be superior to nitric acid.

5.6.5 Filling with molten materials

Most of the methods described above tend to produce short lengths of material inside the tubes, rather than the continuous filling necessary to produce genuine nanowires. Complete filling has proved rather difficult for a number of reasons. In the case of the Ajayan/Iijima experiments using molten lead, filling was impeded by the presence of internal caps traversing the central cavity. As already noted, these barriers can apparently be removed by treatment with liquid oxidants, in particular nitric acid, but end-to-end filling of these opened tubes is still not straightforward. Deposition from solution can never result in complete filling, since the precipitated material will always occupy far less space than the solution. A number of groups have attempted to avoid this problem by treating opened tubes with molten materials. Ajayan and colleagues prepared a sample of tubes by treatment with oxygen or nitric acid and then mixed the sample with V_2O_5 powder (m.p. 690 °C) and heated to a temperature of 750 °C (5.43). This work is discussed further in the next chapter (Section 6.5.2). The Oxford group have shown that treatment of opened nanotubes with molten MoO_3, followed by heating in hydrogen, results in tubes completely filled with single-crystal MoO_2 (5.44). An example is shown in Fig. 5.9.

It should be noted that complete filling of nanotubes has also been achieved using arc-evaporation by Guerret-Piécourt and colleagues (5.24).

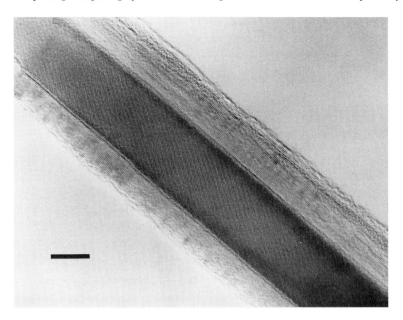

Fig. 5.9. Micrograph showing nanotube completely filled with MoO_2 (5.44). Scale bar 10 nm.

5.6.6 Experiments on capillarity and wetting

In the course of the studies described above, it became clear that some materials entered the opened tubes much more readily than others. Whether or not a liquid will enter the central core of a nanotube depends to a large extent on the interfacial energy between the two. If the liquid–solid contact angle is less than 90°, liquid will enter the tube spontaneously, while if the angle is greater than 90° it will not. More information on the wetting characteristics of nanotubes would therefore be very useful. The only detailed study which has been carried out in this area is the work of Dujardin, Ebbesen and colleagues of NEC (5.45). These workers studied the wetting of nanotubes with a range of materials in an attempt to determine the 'critical' surface energy, below which wetting would occur. They concluded that this cutoff surface energy lies somewhere between 100 and 200 mN/m. Thus water, with a surface energy of ~ 72 mN/m, would be expected to enter nanotubes spontaneously, as would most organic solvents, which have lower surface energies than water. This is consistent with the observation that aqueous solutions enter opened tubes, as described above.

Some important experiments on capillarity-induced filling of nanotubes have been described by Daniel Ugarte and colleagues (5.46). These workers prepared opened nanotubes by oxidation in air, and then annealed the opened tubes at 2000 °C in order to graphitize any residual amorphous carbon, leaving the tube entrances clear. The tubes were then heated with molten silver nitrate. Interestingly, it was found that only those tubes larger than about 4 nm were filled by this treatment. This is the reverse of what one might expect in the 'macro' world, where narrower tubes normally have a stronger capillary effect. The authors explained their observations in terms of curvature-induced bonding changes in very narrow tubes.

5.6.7 Chemistry and crystallisation in nanotubes

The development of methods for opening and filling carbon nanotubes has provided some fascinating insights into the crystallisation behaviour of materials in confined spaces. In their work on nanotubes filled with lead (5.37), Ajayan and Iijima frequently observed that the filling was disordered, particularly in the narrowest tubes. Occasionally lattice fringes were observed in the fillings, but these did not correspond to any known compound. This appears to suggest that crystallisation is inhibited inside the nanotubes. Similar observations were made in the work of Ajayan *et al.* on tubes filled with V_2O_5 (5.43), where the filling in cavities smaller than about 3 nm appeared to be amorphous. This behaviour has been discussed theoretically by Prasad and Lele (5.47), who have shown that a critical diameter exists below which an amorphous phase is stabilised. The value of this critical diameter depends on the interfacial energies of the encapsulated material.

Where crystals are formed inside nanotubes, a number of distinct types of behaviour have been observed. For example, in some cases the enclosed crystallites form discrete particles, while in other cases they form continuous 'nanorods'. The former behaviour is usually observed when the solution method is used to fill the tubes. Examples can be seen in Figs. 5.10 and 5.11, which show crystals of UO_2 (5.32) and SnO (5.48) crystallites inside nanotubes. In the case of the UO_2 crystals, there is evidence of an epitaxial relationship with the tube walls (see particularly Fig. 5.10(b)). The SnO crystallites shown in Fig. 5.11 seem to be randomly oriented, but apparently display a spiral arrangement inside the tube. Spiralling growth inside nanotubes has also been reported by Guerret-Piécourt *et al.* (5.24). An example of a continuous crystallite, in this case Sm_2O_3, inside a nanotube is shown in Fig. 5.12.

Filling with molten materials usually results in continuous crystallites rather than discrete particles. The micrograph of a tube filled with MoO_2 shown in Fig. 5.9 is an example. In such completely filled tubes, an orientation relationship with the tube walls is usually present.

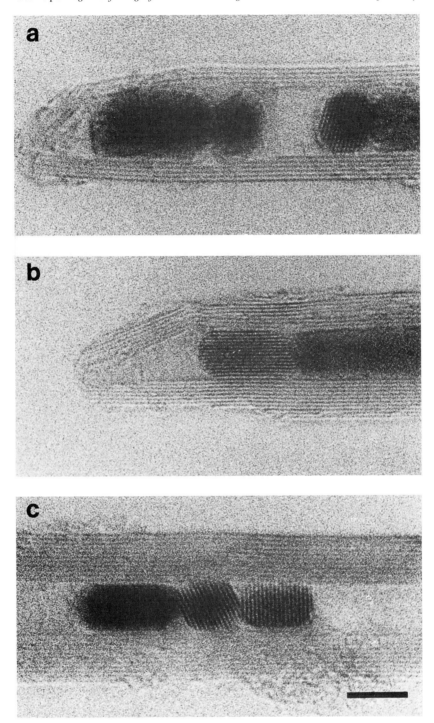

Fig. 5.10. (a)–(c) Nanotubes filled with uranium dioxide, prepared by treatment of tube sample with a nitric acid solution of uranyl nitrate (5.32). Scale bar 5 nm.

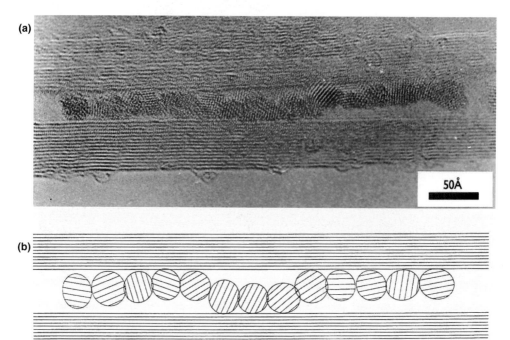

Fig. 5.11. (a) Randomly oriented crystallites of SnO apparently displaying spiral configuration inside nanotube. (b) Schematic representation of spiralling crystallites. (5.48).

There have now been a number of studies in which chemistry has been carried out inside nanotubes. As noted above, Tsang and colleagues prepared tubes containing crystalline nickel oxide by adding nickel nitrate to the nitric acid used in the oxidation treatment. Further treatment with hydrogen at 400 °C reduced the oxide to nickel metal, showing that chemical reactions could be effected inside the opened tubes. A micrograph showing a tube containing metallic nickel is shown in Fig. 5.13. The preparation of MoO_2-filled tubes, described above, also involved reduction of material inside nanotubes. In their work on tubes filled with silver nitrate, Daniel Ugarte and colleagues also performed chemistry inside the tubes (5.46). By focusing the electron beam on the filled tubes, they decomposed the $AgNO_3$ into Ag, NO_2 and O_2. As a result, the continuous crystallites of $AgNO_3$ broke up into discrete Ag particles and erosion of the inner nanotube layers by the evolved gases was observed.

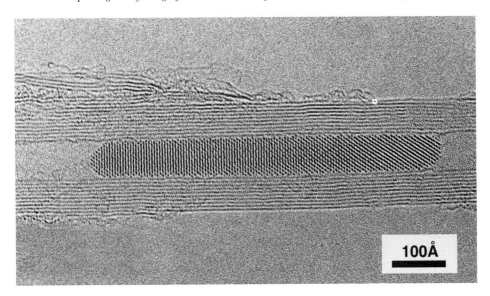

Fig. 5.12. Samarium oxide (Sm_2O_3) crystallite inside nanotube (5.49). Prepared by treatment of closed tubes with $Sm(NO_3)_3/HNO_3$.

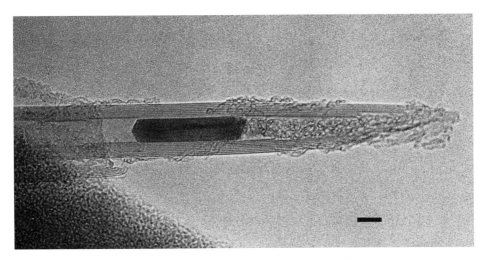

Fig. 5.13. Nanotube filled with metallic nickel prepared by hydrogen reduction of a NiO precursor (5.32). Scale bar 5 nm.

5.6.8 Biological molecules in nanotubes

Following their work on opening nanotubes using nitric acid, Malcolm Green and Edman Tsang joined with bioinorganic chemists from Oxford and from Birkbeck College, London, in an attempt to introduce biological molecules into opened tubes. Their success in doing so, described in September 1995 (5.50), opens up a fascinating new avenue in nanotube research. The biological materials they chose for the first experiments were small enzyme molecules containing heavy elements, which would produce strong contrast in the electron microscope. The technique was simply to suspend samples of opened nanotubes in aqueous solutions of the proteins for 24 hours and then evaporate off the volatile water under reduced pressure. In this way they successfully introduced the enzymes Zn_2Cd_5-metallothionein, cytochrome c_3 and β-lactamase I into the nanotubes' central cavities in high yield. An interesting aspect of this work was that the protein molecules appeared to be undamaged by the TEM beam, possibly having been protected by the electrically conducting nanotubes.

It is conceivable that nanotubes filled or coated with enzymes might be used as electrodes for biosensors. If an enzyme was used which could take part in electron-transfer reactions, this could accept electrons from a substrate and pass them along a conducting nanotube to an electrode surface. Nanotubes are so small that, in principle, an array of them containing different enzymes could be situated on a single microelectrode enabling many analyses to be made simultaneously.

In 1997, the same group described the immobilisation of labelled DNA oligomer on carbon nanotubes and nanoparticles (5.51). The heavy atoms platinum and iodine were used as labels in order to facilitate the imaging of the oligomer by TEM. Both closed and opened nanotubes were treated with the oligonucleotides, and HREM showed the treated tubes to be covered with an amorphous layer 0.7–1.3 nm thick which the authors interpreted as representing the heavy atom labels attached to the oligomer. Like the enzyme-treated nanotubes, these DNA coated tubes may also have potential as biosensors.

5.7 Filling of single-walled nanotubes

Up to this point, we have considered exclusively the filling of multiwalled nanotubes and nanoparticles. Since 1997, a number of papers have appeared which describe the filling of single-walled nanotubes. The first such report involved filling SWNTs with hydrogen, as described in the next section. This was followed by the demonstration by Jeremy Sloan and colleagues from Oxford that SWNTs can be opened and filled using techniques similar to those

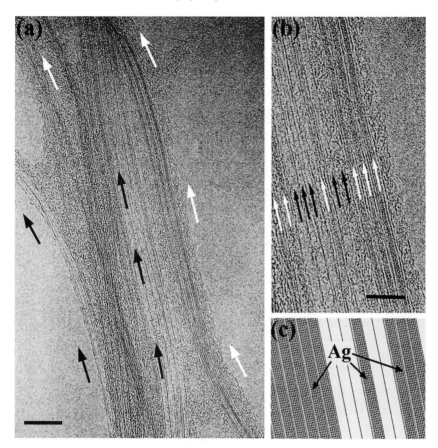

Fig. 5.14. (a) Micrograph illustrating filling of single-walled nanotubes with silver, scale bar 9 nm. (b) Enlargement of (a), scale bar 5 nm. (c) Schematic representation of (b) (5.53).

employed for MWNTs. In an initial study they introduced ruthenium crystallites and other materials into the tubes using a solution method (5.52). Subsequent work showed that almost complete filling of SWNTs could be achieved using molten materials (5.53). Figure 5.14 shows some examples of SWNTs filled with silver, prepared by treating the tubes with AgBr followed by photolytic reduction. This outstanding work opens the way to the use of SWNTs as templates for the synthesis of nanowires with well-defined diameters and to further studies of crystallisation in extremely confined spaces.

At the end of 1998, a group from the University of Pennsylvania showed that SWNTs produced by the Rice pulsed laser vaporisation technique sometimes contain small fullerenes with diameters close to that of C_{60} (5.54). One of their amazing micrographs showing a chain of fullerenes packed inside a tube like

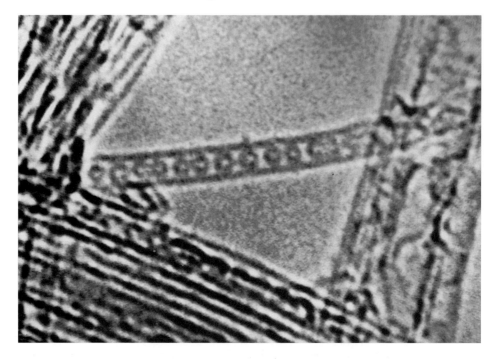

Fig. 5.15. Fullerenes in single-walled nanotube (5.54).

peas in a pod is shown in Fig. 5.15. Quite how the fullerenes came to be encapsulated in this way is a mystery, but further studies might yield clues to the mechanism of tube formation.

5.8 Storing gases in nanotubes

In early 1997, a US group demonstrated the condensation of hydrogen to high density inside single-walled nanotubes (5.55). The high uptakes observed suggest that single-walled tubes might make useful candidates as hydrogen-storage materials. The experimental method used by these workers involved exposing the tube-containing material to hydrogen at 300 Torr, cooling to 90 K and then using temperature programmed desorption (TPD) spectroscopy to observe the desorption behaviour. The as-prepared SWNT soot displayed desorption behaviour similar to that of an activated carbon sample. However, soot which had been heated in vacuum at 970 K showed an extra desorption peak which was apparently consistent with adsorption of hydrogen within the cavities of the tubes. The authors suggested that the tubes had been opened by the heat treatment in vacuum, and thus made accessible to the hydrogen.

A short time later, Australian workers described experiments in which argon was trapped inside graphitic tubes (5.56). These tubes were prepared by catalytic reduction of CO_2 and were much larger than typical carbon nanotubes, having diameters in the range 20 to 150 nm. The argon was introduced by hot isostatically pressing the carbon material for 48 hours at 650 °C under a pressure of 170 megapascals, and could be detected inside the tubes using energy dispersive X-ray spectroscopy. The argon appeared to enter the tubes through defects in the relatively imperfect structures, and may have become sealed inside as a result of amorphization during the HIPing process. The pressure inside the tubes was estimated at 60 megapascals, and appeared to change little over several months at room temperature, so it seems possible to store gases inside these tubes for long periods. The authors of this work entitled their paper 'The world's smallest gas cylinders', although this distinction perhaps belongs to the single-walled nanotubes employed by the American group.

In view of these promising early results, it seems that the encapsulation of gases inside carbon nanotubes and nanoparticles is a field which deserves further attention. One area which may be particularly interesting, and which is alluded to by the Australian workers, is the encapsulation of radioactive gases.

5.9 Discussion

Research into the filling of carbon nanoparticles and nanotubes has been driven both by the potential for applications and by the opportunity for exploring new science. As far as applications are concerned, two areas which appear to have particular potential in the short term are the encapsulation of magnetic materials and the encapsulation of radioisotopes. For both applications, completely sealed nanocapsules are required, so research in these areas has generally involved using the arc-evaporation method to produce filled nanoparticles, rather than the opening and filling of nanotubes. Initial experiments on nanoparticles filled with magnetic and nuclear materials were discussed in Section 5.4. While the early results look promising, there are clearly some obstacles to overcome before commercialisation can be envisaged. In particular, scale-up of the arc-evaporation process may prove difficult. It should also be noted that the arc-evaporation synthesis of filled nanoparticles often results in the formation of carbides, which may not always be desirable. Other possible applications of the arc-evaporation using modified electrodes, such as the synthesis of quantum dots, have not yet been explored in detail.

The modified electrode technique is somewhat limited, in that it can only be applied to those materials which can survive the extreme conditions of the arc. Chemical techniques of opening the nanocapsules offer a way of encapsulating more delicate materials. Experiments in this area have usually involved nanotubes rather than nanoparticles, and a whole range of materials have been encapsulated, ranging from oxides to biomolecules. Here, the motivation has been primarily scientific, and a number of new insights into the behaviour of materials in enclosed nanospaces have been obtained. Nevertheless, there is clearly some potential for applications of opened and filled nanotubes. One possibility is to use nanotubes as moulds for quantum wires or 'nanorods', which may have interesting mechanical or physical properties (see also Section 7.4). Another idea is to use nanotubes containing biological molecules as nanosensors, as discussed in Section 5.6.8. However, such applications probably lie well in the future.

References

(5.1) J. R. Heath, S. C. O'Brien, Q. Zhang, Y. Liu, R. F. Curl, H. W. Kroto, F. K. Tittel, and R. E. Smalley, 'Lanthanum complexes of spheroidal carbon shells', *J. Amer. Chem. Soc.*, **107**, 7779 (1985).

(5.2) D. S. Bethune, R. D. Johnson, J. R. Salem, M. S. de Vries, and C. S. Yannoni, 'Atoms in carbon cages: the structure and properties of endohedral fullerenes', *Nature*, **366**, 123 (1993).

(5.3) W. Sliwa, 'Metallofullerenes', *Trans. Met. Chem.*, **21**, 583 (1996).

(5.4) P. R. Birkett, A. G. Avent, A. D. Darwish, H. W. Kroto, R. Taylor and D. R. M. Walton, 'Holey fullerenes – a bis-lactone derivative of [70] fullerene with an 11-atom orifice', *J. Chem. Soc., Chem. Commun.*, 2489 (1996).

(5.5) Y. Chai, T. Guo, C. M. Jin, R. E. Haufler, L. P. F. Chibante, J. Fure, L. H. Wang, J. M. Alford and R. E. Smalley 'Fullerenes with metals inside', *J. Phys. Chem.*, **95**, 7564 (1991).

(5.6) D. E. Manolopoulos and P. W. Fowler, 'Structural proposals for endohedral metal fullerene complexes', *Chem. Phys. Lett.*, **187**, 1 (1991).

(5.7) R. D. Johnson, M. S. de Vries, J. R. Salem, M. S. de Vries, D. S. Bethune and C. S. Yannoni, 'Electron paramagnetic resonance studies of lanthanum-containing C_{82}', *Nature*, **355**, 239 (1992).

(5.8) R. S. Ruoff, D. C. Lorents, B. Chan, R. Malhotra and S. Subramoney, 'Single crystal metals encapsulated in carbon nanoparticles', *Science*, **259**, 346 (1993).

(5.9) M. Tomita, Y. Saito and T. Hayashi, 'LaC_2 encapsulated in graphite nanoparticle', *Jap. J. Appl. Phys.*, **32**, L280 (1993).

(5.10) Y. Saito in *Carbon nanotubes: preparation and properties*, ed. T. W. Ebbesen, CRC Press, Boca Raton, 1997, p. 249.

(5.11) Y. Saito, 'Nanoparticles and filled nanocapsules', *Carbon*, **33**, 979 (1995).

(5.12) S. Seraphin, D. Zhou and J. Jiao, 'Filling the carbon nanocages', *J. Appl. Phys.*, **80**, 2097 (1996).

(5.13) Y. Murakami, T. Shibata, K. Okuyama, T. Arai, H. Suematsu, and Y. Yoshida, 'Structural, magnetic and superconducting properties of graphite nanotubes and their encapsulation compounds', *J. Phys. Chem. Solids*, **54**, 1861 (1993).

(5.14) Y. Saito, T. Yoshikawa, M. Okuda, M. Ohkohchi, Y. Ando, A. Kasuya and Y. Nishina, 'Synthesis and electron-beam incision of carbon nanocapsules encaging YC_2', *Chem. Phys. Lett.*, **209**, 72 (1993).

(5.15) Y. Saito, T. Yoshikawa, M. Okuda, N. Fujimoto, K. Sumiyama, K. Suzuki, A. Kasuya, and Y. Nishina, 'Carbon nanocapsules encaging metals and carbides', *J. Phys. Chem. Solids*, **54**, 1849 (1993).

(5.16) Y. Saito in *Recent advances in the chemistry and physics of fullerenes and related materials, Vol.* 1, ed. K. M. Kadish and R. S. Ruoff, Electrochemical Society, Pennington, New Jersey, 1994, p. 1419.

(5.17) Y. Saito, K. Nishikubo, K. Kawabata and T. Matsumoto, 'Carbon nanocapsules and single-layered nanotubes produced with platinum-group metals (Ru, Rh, Pd, Os, Ir, Pt) by arc-discharge', *J. Appl. Phys.*, **80**, 3062 (1996).

(5.18) S. Seraphin, D. Zhou, J. Jiao, J. C. Withers and R. Loutfy, 'Yttrium carbide in nanotubes', *Nature*, **362**, 503 (1993).

(5.19) J. Jiao, S. Seraphin, X. Wang and J. C. Withers, 'Preparation and properties of ferromagnetic carbon-coated Fe, Co and Ni nanoparticles', *J. Appl. Phys.*, **80**, 103 (1996).

(5.20) S. A. Majetich, J. O. Artman, M. E. McHenry, N. T. Nuhfer and S. W. Staley, 'Preparation and properties of carbon-coated magnetic nanocrystallites', *Phys. Rev. B*, **48**, 16 845 (1993).

(5.21) M. E. McHenry, S. A. Majetich, J. O. Artman, M. DeGraef and S. W. Staley, 'Superparamagnetism in carbon-coated Co particles produced by the Kratschmer carbon arc process', *Phys. Rev. B*, **49**, 11 358 (1994).

(5.22) J. H. Scott and S. A. Majetich, 'Morphology, structure, and growth of nanoparticles produced in a carbon arc', *Phys. Rev. B*, **52**, 12 564 (1995).

(5.23) V. P. Dravid, J. Host, M. H. Teng, B. Elliott, J. Hwang, D. L. Johnson, T. O. Mason and J. R. Weertman, 'Controlled-size nanocapsules', *Nature*, **374**, 602, (1995).

(5.24) C. Guerret-Piécourt, Y. Le Bouar, A. Loiseau and H. Pascard, 'Relation between metal electronic structure and morphology of metal compounds inside carbon nanotubes', *Nature*, **372**, 761 (1994).

(5.25) P. J. F. Harris and S. C. Tsang, 'A simple technique for the synthesis of filled carbon nanoparticles', *Chem. Phys. Lett.*, **293**, 53 (1998).

(5.26) P. J. F. Harris and S. C. Tsang, 'Encapsulating uranium in carbon nanoparticles using a new technique', *Carbon*, **36**, 1859 (1998).

(5.27) Y. Yosida, 'Synthesis of CeC_2 crystals encapsulated within gigantic super fullerenes', *Appl. Phys. Lett.*, **62**, 3447 (1993).

(5.28) T. Guo, M. D. Diener, Y. Chai, M. J. Alford, R. E. Haufler, S. M. McClure, Y. Ohno, J. H. Weaver, G. E. Scuseria and R. E. Smalley, 'Uranium stabilisation of C_{28}: a tetravalent fullerene', *Science*, **257**, 1661 (1992).

(5.29) K. Kikuchi, K. Kobayashi, K. Sueki, S. Suzuki, H. Nakahara, Y. Achiba, K. Tomura and M. Katada, 'Encapsulation of radioactive [159]Gd and [161]Tb in fullerene cages', *J. Amer. Chem. Soc.*, **116**, 9775 (1994).

(5.30) H. Funasaka, K. Sugiyama, K. Yamamoto and T. Takahashi, 'Synthesis of actinide carbides encapsulated within carbon nanoparticles', *J. Appl. Phys.*, **78**, 5320 (1995).

(5.31) E. Pasqualini, P. Adelfang and M. N. Regueiro, 'Carbon nanoencapsulation of uranium dicarbide', *J. Nucl. Mat.*, **231**, 173 (1996).

(5.32) S. C. Tsang, Y. K. Chen, P. J. F. Harris and M. L. H. Green, 'A simple chemical method of opening carbon nanotubes', *Nature*, **372**, 159 (1994).

(5.33) A. Kasuya, H. Takahashi, Y. Saito, T. Mitsugashira, T. Shibayama, Y. Shiokawa, I. Satoh, M. Fukushima and Y. Nishima, 'Neutron irradiation on carbon nanocapsules', *Mat. Sci. Eng. A*, **217**, 50 (1996).

(5.34) W. M. Burch, P. J. Sullivan and C. J. McLaren, 'Technegas – a new ventilation agent for lung scanning', *Nucl. Med. Commun.*, **7**, 865 (1986).

(5.35) L. Dayton and J. Johnson, 'Buckyballs "covered by old patent"', *New Scientist*, 6 July 1991, p. 26.

(5.36) T. J. Senden, K. H. Moock, J. FitzGerald, W. M. Burch, R. J. Browitt, C. D. Ling and G. A. Heath, 'The physical and chemical nature of Technegas', *J. Nucl. Med.*, **38**, 1328 (1997).

(5.37) P. M. Ajayan and S. Iijima, 'Capillarity-induced filling of carbon nanotubes', *Nature*, **361**, 333 (1993).

(5.38) M. R. Pederson and J. Q. Broughton, 'Nanocapillarity in fullerene tubules', *Phys. Rev. Lett.*, **69**, 2689 (1992).

(5.39) S. C. Tsang, P. J. F. Harris and M. L. H. Green, 'Thinning and opening of carbon nanotubes by oxidation using carbon dioxide', *Nature*, **362**, 520 (1993).

(5.40) P. M. Ajayan, T. W. Ebbesen, T. Ichihashi, S. Iijima, K. Tanigaki and H. Hiura, 'Opening carbon nanotubes with oxygen and implications for filling', *Nature*, **362**, 522 (1993).

(5.41) H. Hiura, T. W. Ebbesen and K. Tanigaki, 'Opening and purification of carbon nanotubes in high yields', *Advanced Materials*, **7**, 275 (1995).

(5.42) K. C. Hwang, 'Efficient cleavage of carbon graphene layers by oxidants', *J. Chem. Soc., Chem. Commun.*, 173 (1995).

(5.43) P. M. Ajayan, O. Stephan, P. Redlich and C. Colliex, 'Carbon nanotubes as removable templates for metal oxide nanocomposites and nanostructures', *Nature*, **375**, 564 (1995).

(5.44) Y. K. Chen, M. L. H. Green and S. C. Tsang, 'Synthesis of carbon nanotubes filled with long continuous crystals of molybdenum oxides', *J. Chem. Soc., Chem. Commun.*, 2489 (1996).

(5.45) E. Dujardin, T. W. Ebbesen, H. Hiura and K. Tanigaki, 'Capillarity and wetting of carbon nanotubes', *Science*, **265**, 1850 (1994).

(5.46) D. Ugarte, A. Châtelain and W. A. de Heer, 'Nanocapillarity and chemistry in carbon nanotubes', *Science*, **274**, 1897 (1996).

(5.47) R. Prasad and S. Lele, 'Stabilization of the amorphous phase inside carbon nanotubes – solidification in a constrained geometry', *Philos. Mag. Lett.*, **70**, 357 (1994).

(5.48) J. Sloan, J. Cook, R. Heesom, M. L. H. Green and J. L. Hutchison, 'The encapsulation and in situ rearrangement of polycrystalline SnO inside carbon nanotubes', *J. Cryst. Growth*, **173**, 81 (1997).

(5.49) J. Sloan, J. Cook, M. L. H. Green, J. L. Hutchison and R. Tenne, 'Crystallisation inside fullerene related structures', *J. Mater. Chem.*, **7**, 1089 (1997).

(5.50) S. C. Tsang, J. J. Davis, M. L. H. Green, H. A. O. Hill, Y. C. Leung and P. J. Sadler, 'Immobilisation of small proteins in carbon nanotubes: High resolution transmission electron microscopy study and catalytic activity', *J. Chem. Soc., Chem. Commun.*, 1803 (1995).

(5.51) S. C. Tsang, Z. Guo, Y. K. Chen, M. L. H. Green, H. A. O. Hill, T. W.

Hambley and P. J. Sadler, 'Immobilisation of platinised and iodinated oligonucleotides on carbon nanotubes', *Angew. Chem. Int. Ed.*, **36**, 2198 (1997).

(5.52) J. Sloan, J. Hammer, M. Zwiefka-Sibley and M. L. H. Green, 'The opening and filling of single walled carbon nanotubes (SWTs)', *J. Chem. Soc., Chem. Commun.*, 347 (1998).

(5.53) J. Sloan, D. M. Wright, H. G. Woo, S. Bailey, G. Brown, A. P. E. York, K. S. Coleman, J. L. Hutchison and M. L. H. Green, 'Surface wetting, capillarity and silver nanowire formation observed in single-walled carbon nanotubes', *J. Chem. Soc., Chem. Commun.*, 699 (1999).

(5.54) B. W. Smith, M. Monthioux and D. E. Luzzi, 'Encapsulated C_{60} in carbon nanotubes', *Nature*, **396**, 323 (1998).

(5.55) A. C. Dillon, K. M. Jones, T. A. Bekkedahl, C. H. Kiang, D. S. Bethune and M. J. Heben, 'Storage of hydrogen in single-walled carbon nanotubes', *Nature*, **386**, 377, (1997).

(5.56) G. E. Gadd, M. Blackford, D. Moricca, N. Webb, P. J. Evans, A. M. Smith, G. Jacobsen, S. Leung, A. Day and Q. Hua, 'The world's smallest gas cylinders', *Science*, **277**, 933 (1997).

6

The ultimate carbon fibre?

The mechanical properties of carbon nanotubes

Finally, under a scanning electron microscope, the last struts would
be replaced by the last set of sub-sub-miniature Tensegrity Masts,
down at the order of size where we speak of an atom's diameter. . . .
The last tensile wires will be simply the chemical bonds.
>   Hugh Kenner, *Bucky: A Guided Tour of Buckminster Fuller*

The sp^2 carbon–carbon bond in the basal plane of graphite is the strongest of
all chemical bonds, but the weakness of the interplanar forces means that
ordinary graphite is of little value as a structural material. One way in which
the great strength of the sp^2 bonds can be exploited practically is through the
use of fibres in which all the basal planes run approximately parallel to the axis.
Carbon fibres with this structure have been produced for many years, and are
now widely used as components of composite materials in applications
ranging from golf clubs to military aircraft (6.1–6.4). These conventional
carbon fibres, which are prepared by the controlled pyrolysis of various
precursors, possess very high stiffness and strength, but their properties still fall
well short of the theoretical maxima owing to structural imperfections. When
fullerene-related carbon nanotubes were discovered in 1991, their highly perfect structure prompted speculation about their potential as 'super-strong'
carbon fibres. Early theoretical work by David Tomanek and his colleagues
from Michigan State University (6.5) confirmed that nanotubes should possess
exceptional mechanical properties, and in fact should have a higher rigidity
than that of any known material. Experimental confirmation of this has
proved more difficult, for obvious reasons: the stress–strain behaviour of
structures with diameters of the order of 10 nm clearly cannot be tested using
conventional techniques. However, elegant work by Michael Treacy and
Thomas Ebbesen of NEC and Murray Gibson of the University of Illinois has
enabled measurements of the intrinsic thermal vibrations of multiwalled tubes

to be made using TEM (6.6), and these confirm that the tubes have exceptionally high elastic moduli. Direct measurements using atomic force microscopy (AFM) have provided further evidence of the outstanding mechanical properties of individual nanotubes. This chapter summarises the theoretical and experimental work which has been carried out on the mechanical properties of carbon nanotubes, together with the limited amount of work which has been done on incorporating nanotubes into matrices. The use of carbon nanotubes as tips for scanning probe microscopy is also reviewed. To begin, an outline is given of the preparation, properties and applications of more well-established types of carbon fibres: 'conventional' carbon fibres, prepared by pyrolysis, graphite whiskers and catalytically grown filaments.

6.1 Conventional carbon fibres

The first attempts to produce carbon fibres by the pyrolysis of fibrous organic material were made by Thomas Edison in the USA and Joseph Swan in England over a century ago (6.7). The application of these strands was not structural, but as filaments in the first incandescent electric lamps. After trying a huge range of organic precursors, Edison found that carbonized cotton thread gave the best results. Swan had considerable success preparing filaments by carbonizing threads extruded from regenerated cellulose. Ultimately carbon filaments were replaced by the much longer-lasting tungsten filaments in use today.

There was little further interest in carbon fibres until the development of jet propulsion in the 1950s led to a demand for low density, high stiffness materials. This prompted a number of groups in the aircraft industry to explore new synthetic routes to carbon fibres. Particularly notable was the work of William Watt and colleagues of the Royal Aircraft Establishment, Farnborough. Like Swan many years earlier, Watt experimented with carbonized cellulose, but soon found that the synthetic polymer polyacrylonitrile (PAN) was a more suitable precursor. The process he used to carbonize the polymer fibres involved firstly heating them under tension in an inert atmosphere at 250 °C, then oxidising in air at 600 °C to produce a linked-ring structure which could then be readily graphitized by high-temperature heat treatment. The mechanical properties of the fibres depended to a large extent on the temperature of this final heat treatment.

A vast amount of research on carbon fibres derived from organic materials has been carried out since the work of Watt, and a variety of different precursors have been tried, with varying results. An important development was the introduction by Union Carbide in the 1970s of fibres derived from petroleum or coal-tar pitch. When certain pitches are heated to $\sim 375\,^{\circ}\text{C}$ a highly oriented anisotropic liquid crystalline state, known as mesophase, results (see also p. 260). This material can be spun into fibres with highly preferred orientations, which can then be carbonized and graphitized by high-temperature heat treatment.

There are considerable differences between the microstructures and mechanical properties of carbon fibres derived from polyacrylonitrile and those prepared from pitch. These result from the fact that PAN is a non-graphitizing carbon while pitch is graphitizing (6.8). Thus, PAN-derived fibres only contain very small graphitic domains, and have a large number of voids, giving them a relatively low density. This lack of extended structure makes ex-PAN fibres relatively insensitive to flaws, giving them exceptionally high strengths. Since pitch is a graphitizing carbon, pitch-derived fibres have a much more perfect graphitic structure, which gives them a higher density and results in higher elastic moduli than for the PAN fibres. This is illustrated in Fig. 6.1, which shows stress–strain curves for ex-PAN and ex-pitch fibres which have been graphitized at various temperatures. The curves for PAN-derived fibres imply Young's moduli in the approximate range 250–300 GPa, while the pitch-derived fibres have moduli of the order of 300–800 GPa. Also included in the graph is the stress–strain curve corresponding to the in-plane Young's modulus, C_{11}, of single-crystal graphite. This has a value of 1060 GPa, which represents the maximum theoretical modulus for a carbon fibre.

The industrial use of carbon fibres has increased rapidly since the mid-1960s, although most applications remain specialised. The relatively high cost of carbon fibre production has inhibited their widespread use in mass market products like automobiles.

6.2 Graphite whiskers

At about the same time that Watt and others were preparing carbon fibres by pyrolysis, a quite different kind of carbon filament was discovered by Roger Bacon, a physicist working at the Union Carbide Corporation in Cleveland, Ohio. These new fibres, which Bacon named graphite whiskers, were grown in a DC carbon arc under a high pressure of inert gas (6.9). The whiskers were found at the interior of a large 'boule' of carbon which was observed to grow like a stalagmite on the negative lower electrode (the electrodes having been arranged in a vertical configuration). Electron microscopy showed the whiskers to be far more perfect in structure than the pyrolytically produced carbon fibres. The structure appeared to be that of a scroll made up of an essentially *continuous* graphitic structure, as shown in Fig. 6.2. As a result of

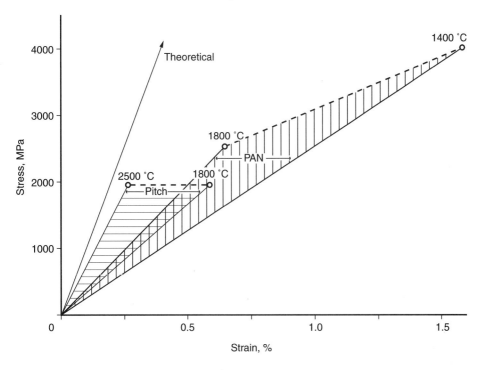

Fig. 6.1. Stress–strain curves for carbon fibres derived from polyacrylonitrile (PAN) and from pitch. Theoretical line corresponds to in-plane Young's modulus of graphite.

this high degree of perfection, the whiskers had exceptional mechanical properties: Bacon reported tensile strengths up to 20 GPa and Young's moduli of approximately 800 GPa, both of which exceed the values for most organically derived carbon fibres. However, the high cost and low yield of the synthesis process has meant that graphite whiskers have never become a commercial product.

The method used to prepare graphite whiskers is clearly very similar to that used for carbon nanotubes, and there are obvious similarities between the structures of the two types of fibre. Therefore, when nanotubes were discovered it was natural to ask whether they were not simply smaller versions of graphite whiskers. The answer is almost certainly no, for a number of reasons. Firstly, the observed size ranges are quite different. In samples prepared using the standard Ebbesen–Ajayan method, the largest nanotubes are approximately 100 nm, or 0.1 μm, in diameter, and generally no more than a few micrometres in length. Graphite whiskers, on the other hand, typically have diameters of the order of 5 μm, and can be up to 3 cm in length. Secondly, graphite whiskers are not closed structures like nanotubes: electron micrographs show the ends to be

Fig. 6.2. Sketch of scroll-like structure of graphite whiskers (6.9).

irregular and frayed rather than capped. Thirdly, graphite whiskers have a
scrolled, rather than a nested, coaxial structure. This was demonstrated in an
ingenious experiment in which Bacon 'unrolled' a whisker to produce graphite
sheets which were much larger than the whisker diameter (6.9).

6.3 Catalytically grown carbon fibres

The vapour-phase catalytic synthesis of carbon nanotubes was described in
Chapter 2 (Section 2.3). There is considerable interest in using catalytically
grown fibres in structural composites, and a substantial amount of work has
been done on their mechanical properties (6.2, 6.10). Measurements have been
made on fibres with a range of diameters, and the properties were found to
improve with falling diameter. Thus, for fibres with a diameter of 40 μm, tensile
strengths of the order of 1 GPa and Young's moduli of approximately 170 GPa
are observed. These figures rise to $\sim$ 2.5 GPa and 350 GPa respectively for
fibres with diameters less than 10 μm. The poorer properties of the thicker
fibres are attributed to a faster growth rate, leading to lower crystallinity.
These excellent properties, in fibres produced using relatively low temperatures ($\sim$ 1100 °C), make catalytically-grown fibres an attractive commercial
prospect.

6.4 Mechanical properties of carbon nanotubes

6.4.1 Theoretical predictions

Before discussing detailed calculations of the mechanical properties of nanotubes, we can carry out some simple calculations to illustrate the relationship between the diameter of a nanotube and its stiffness. Consider first a tube with an inner diameter of 1 nm. We assume a wall thickness of 0.34 nm, so the outer diameter is 1.68 nm and the cross-sectional area is 1.43×10^{-18} m^2. If we now apply a tensile load of 100 nN to the tube, this results in a stress of $\sim$ 70 GPa. The corresponding strain, assuming a Young's modulus of 1060 GPa, is approximately 6.6%. Now consider a tube with an inner diameter of 10.0 nm and an outer diameter of 10.68 nm. In this case a tensile load of 100 nN results in a stress of 9.05 GPa, and a strain of about 0.85%. These figures clearly demonstrate the way in which stiffness increases with tube diameter. This is consistent with the observation that single-walled nanotubes, with diameters typically of the order of 1 nm, are usually curly while multiwalled tubes tend to be straight.

In the simple calculation given above, it was assumed that the Young's modulus for nanotubes will be equal to that for a graphene sheet, i.e. 1060 GPa. This can only be an approximation, and a number of groups have attempted to calculate the modulus for nanotubes with various diameters and structures, using a variety of approaches. As already noted, some of the earliest theoretical work in this area was carried out by the Tomanek group from Michigan. This group employed the Keating potential to determine the structural rigidity of short, single-walled (5,5) nanotubes, containing 100, 200 and 400 atoms (6.5). They found exceptional rigidities for these tubes, which implied they should be about an order of magnitude stiffer than an iridium beam of comparable diameter. Tomanek and colleagues did not calculate values for the Young's modulus of the tubes because of the difficulty of defining the layer thickness. However, other authors have suggested that the Tomanek results imply a Young's modulus in the range 1500–5000 GPa (6.6). These values are higher than the accepted value for a graphene sheet, which seems illogical, although it is possible that this accepted value may be in error. More recently, Jian Ping Lu has described tight-binding calculations on single-walled tubules with diameters from 0.34 nm to 13.5 nm, and found a Young's modulus of 970 GPa (6.11). This is close to the modulus for a graphene sheet, and was found to be independent of tube structure or diameter. In similar calculations, Angel Rubio and colleagues found slightly higher values for the Young's moduli (typically 1240 GPa) of tubes with a range of structures and diameters (6.12). Unlike Lu, Rubio *et al.* found that the moduli depended on both tube diameter and structure. It seems that more work will be needed before agreement is reached on this point.

Some of the most extensive theoretical work on the mechanical properties of carbon nanotubes has been carried out by workers from North Carolina State University (e.g. 6.13, 6.14). This group has concentrated chiefly on the behaviour of nanotubes under extreme deformations, using both molecular dynamics and macroscopic structural dynamics approaches. Their simulations have included the effect of nanotube buckling, which is observed experimentally (see next section). Some of their results are illustrated in Fig. 6.3 (6.13). Here, the effect of axial compression on an armchair (7,7) nanotube has been simulated using molecular dynamics. Figure 6.3(a) shows a plot of strain energy vs. longitudinal strain, ε. This shows that at small strains, the strain energy varies with ε^2, as expected from Hooke's Law, but at higher strains a series of discontinuities are observed, and the strain energy curve becomes approximately linear. The discontinuities labelled b–e correspond to the four buckled configurations shown in the simulations. Nanotube bending was also simulated, as illustrated in Fig. 6.4 for the case of a (13,0) zig-zag tube. Again a kink is observed in the strain energy curve, corresponding to the buckled structure shown in Fig. 6.4(b).

The NCSU group did not observe fracture in their simulations of axial compression or bending of nanotubes, but the application of tension does eventually produce fracture. This is illustrated in Fig. 6.5, which shows the stages leading to fracture of a (13,0) zig-zag tube. At the top, the tube is in a highly strained state, with the hexagons elongated in the direction of the tube axis. At a critical point, atomic disorder begins to nucleate, and bonds begin to break. A distorted and unstable neck then forms between the two separating segments of the tube, leading ultimately to the formation of two and then one distinct carbon chains, as shown in the lower two simulations. Surprisingly, this individual chain does not break with further separation of the tube ends, but grows in length as further carbon atoms are added from both sides. This remarkable prediction has not yet been tested experimentally.

6.4.2 *Experimental observations using TEM: qualitative*

Determining the mechanical properties of molecular-scale fibres like carbon nanotubes obviously poses great difficulties, and it is only very recently that measurements on individual tubes have been made. However, general observations using transmission electron microscopy have provided some useful insights into the stiffness and strength of nanotubes, and these will be summarised first.

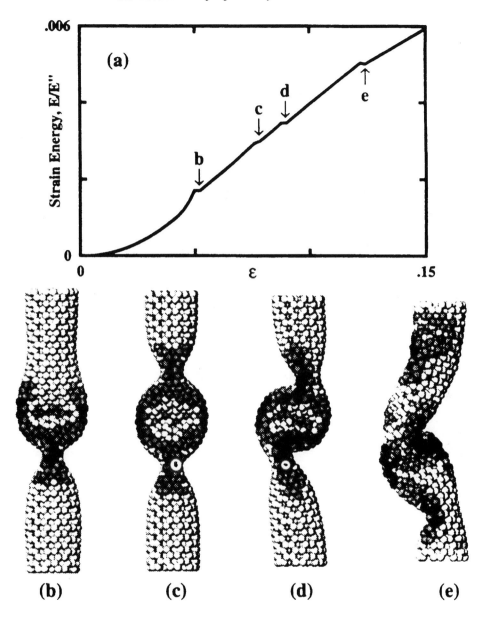

Fig. 6.3. Molecular dynamics simulation of armchair (7,7) nanotube under axial compression, from the work of Yakobson *et al.* (6.13). (a) Plot of strain energy vs. longitudinal strain, ε. (b)–(e) Morphological changes corresponding to singularities in strain energy curves.

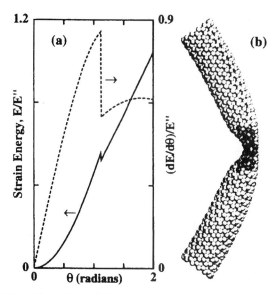

Fig. 6.4. Simulated bending of a (13,0) zig-zag tube. (b) Morphology of tube beyond buckling point (6.13).

One indication that multiwalled nanotubes are reasonably stiff is that they generally appear fairly straight in TEM images (see for example Fig. 1.3). Completely fractured nanotubes are hardly ever observed, even though samples for TEM are sometimes prepared by grinding the material under a solvent in a pestle and mortar. Single-walled nanotubes usually display much more curvature than multiwalled tubes (Figs. 1.8 and 2.17), but this is probably because they are generally extremely thin (as noted in the previous section, stiffness increases with increasing diameter). These observations indicate both a high Young's modulus and a high breaking stress for nanotubes.

Although broken tubes are rarely seen, bent and buckled nanotubes are quite commonly observed (e.g. 6.15–6.18). An example of a buckled multiwalled tube, taken from the work of Despres and co-workers (6.15) is shown in Fig. 6.6. Buckled multiwalled and single-walled tubes have also been described by Iijima and colleagues (6.18), some of whose work is shown in Fig. 6.7. The similarity of the buckling behaviour shown in these images to that observed in macroscopic structures such as rubber tubing is quite striking. The lattice fringes in the buckled regions show enhanced contrast, indicating that they have been flattened during buckling. Some reduction of the interplanar lattice spacing is also observed in the buckled area, while expansion is sometimes seen in regions between the kinks. In some cases, broken graphene layers are observed in the buckled areas, but frequently there is no obvious fracture: a striking example of the flexibility of graphene layers.

Fig. 6.5. Simulation of the effect of axial tension on a (13,0) zig-zag tube (6.14).

Toru Kuzumaki and colleagues from the University of Tokyo (6.20) have developed an ingenious method of observing nanotube buckling directly inside the TEM. This was achieved by gluing the nanotubes to a support film which was then slit near the glued tubes. During observation in the microscope, the film could then be caused to curl due to heating effects, and this curling was sufficient to induce bending and buckling of the supported tubes. Among the tubes which had been bent in this way they often found examples of regularly spaced buckles, as shown in Fig. 6.8. This group have also proposed an atomic model for the buckled structures (6.20).

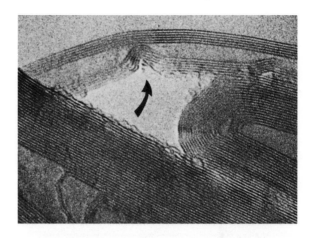

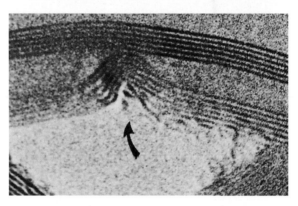

Fig. 6.6. High resolution electron micrographs of buckled nanotube, from the work of Despres and colleagues (6.15).

6.4.3 Experimental observations using TEM: quantitative

The first quantitative TEM measurements of the mechanical properties of nanotubes were carried out by Treacy, Ebbesen and Gibson in 1996 (6.6). Clusters of nanotubes were deposited on TEM grids such that isolated tubes extended for a considerable distance into empty space. The specimens were then placed in a special holder which enabled *in situ* heating to be carried out in the TEM. Images were recorded of a number of individual, freely vibrating, nanotubes at temperatures up to 800 °C. By analysing the mean-square amplitude as a function of temperature it was possible to obtain estimates for the Young's modulus. These ranged from 410 GPa to 4150 GPa, with an average of 1800 GPa. The large spread in values results from uncertainties in estimating the lengths of the anchored tubes, and from the presence of defects in the tube structures.

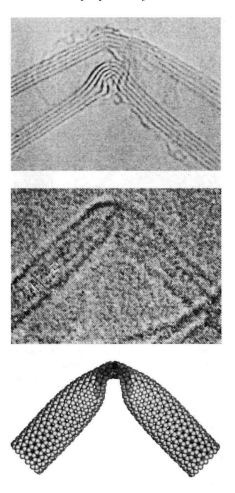

Fig. 6.7. Images by Iijima of buckled multiwalled and single-walled nanotubes, and simulation of structure (6.18, 6.19).

Recent work by Daniel Wagner and colleagues has enabled estimates to be made of the compressive strength of carbon nanotubes (6.21). In this work, multiwalled nanotubes were embedded in an epoxy resin, which was then cut into thin slices in a microtome and examined in a TEM. It was found that many of the tubes had buckled and collapsed following this treatment as a result of stresses arising from polymerisation and from thermal effects associated with the electron beam. Several different types of behaviour were observed. The thicker walled tubes buckled in the way described in the previous section, while thinner walled tubes tended to collapse or possibly fracture. Wagner *et al.* estimated the stresses required to produce buckling or

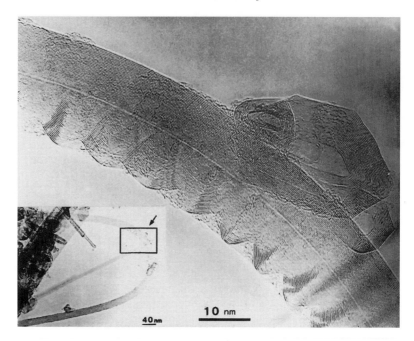

Fig. 6.8. Electron micrograph of deformed nanotube showing periodic buckling (6.20).

collapse, and found values in the approximate range 100–150 GPa. These values imply that carbon nanotubes have compressive strengths at least 100 times higher than those of any other known fibre.

6.4.4 Experimental observations using scanning probe microscopy

While TEM has provided some important insights into the mechanical properties of nanotubes, only scanning probe microscopy enables individual tubes to be manipulated directly. Since about 1996, several groups have used this technique to study the bending and buckling of carbon nanotubes, and made some very striking observations (6.22–6.24). In the first of these, Charles Lieber and colleagues from Harvard described a method of fixing nanofibres at one end, and then determining the bending force of the fibres as a function of displacement (6.22). The method was based on a technique which the same group had used to measure the conductivity of nanotubes (see p. 124). The fibres were firstly dispersed on a single crystal MoS substrate, and then pinned to this surface by depositing square pads of SiO through a mask. This left many fibres with one end trapped under the pads and one end free. The samples were then imaged by AFM, enabling the free nanofibres to be located. Repeated scans of individual fibres then enabled lateral force vs. displacement (F–d) data to be recorded. Scanning could be carried out in such a way that the tip rode over the fibre when a certain applied force was reached, so that the fibre sprang back to its equilibrium position, enabling further scans to be made.

The technique was firstly applied to silicon carbide nanorods, with diameters of the order of 22 nm, which the Lieber group prepared using carbon nanotubes as templates (see p. 229). It was found that the measured lateral force increased linearly as the nanorods were deflected, and that the slopes of the *F–d* curves decreased when measurements were made at increasing distances from the pinning point. Fitting the data to standard beam formulae (taking into account the friction forces between the rods and the substrate) produced values for Young's modulus in the range 610 GPa–660 GPa. These results agreed well with the theoretical figure of 600 GPa for single-crystal SiC. At large deflections the SiC nanorods fractured.

Lieber and colleagues then applied the same technique to multiwalled carbon nanotubes. Again, linear *F–d* curves were observed, at least for small deflections, and the results implied a value of ∼ 1.28 TPa for the Young's modulus. As in some of the other experimental studies, this exceeds the accepted value for a graphene sheet, for reasons which are not completely understood. Unlike the SiC nanorods, the carbon nanotubes could accommodate large deflections without breaking. However, for deflections larger than 10° an abrupt change in the slope of the *F–d* curve was seen. This was attributed to elastic buckling of the kind seen in TEM studies. Support for this view was provided by images of tubes supported on high friction substrates. On such supports the tubes were trapped in highly deflected configurations, and AFM images showed height increases at the most highly curved points.

In a paper published shortly after the Harvard work, Richard Superfine and colleagues from the University of North Carolina at Chapel Hill also described AFM studies of carbon nanotubes subjected to large bending stresses (6.23). For these experiments the tubes were not fixed at one end, but were supported on mica. In many cases the friction was sufficient to pin the tube in a strained configuration for imaging. Again it was found that the tubes could be bent repeatedly through large angles without fracturing. Figure 6.9 shows an individual tube which has been manipulated into two different configurations. Two types of behaviour were identified: small, regularly spaced buckles of the kind seen in TEM studies, and sharper bends involving greater deformations. Examples of both can be seen in the two AFM images. The regularly spaced buckles were found to have characteristic intervals, and disappeared when the tubes were straightened. This suggested that the buckles were reversible and intrinsic features of the nanotubes and did not result from defects. Similar periodic buckling has been observed by TEM, as noted above. The sharper

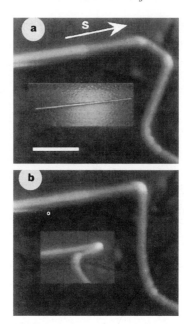

Fig. 6.9. (a), (b) Atomic force microscope images of carbon nanotubes on mica, showing bending and buckling induced by the tip (6.23). The scale bar in (a) is 300 nm.

bends appeared to be associated with permanent defects, although in some cases tubes could be subjected to large deformations without sustaining obvious damage. These experiments provide further evidence of the extraordinary resilience of carbon nanotubes.

The most graphic demonstration of the way in which carbon nanotubes can be manipulated by AFM was provided by Phaedon Avouris and co-workers of IBM in early 1998 (6.24). These workers deposited nanotubes on a hydrogen-passivated Si(100) surface. The strong interaction between the tubes and this surface enabled a range of different kinds of operation to be performed on the tubes, and various controlled manipulations of the tubes were demonstrated. One such sequence, illustrated in Fig. 6.10 involved the creation of a Greek letter Θ. This kind of manipulation might represent one way in which nanotubes could be used to make connections in nanoelectronic circuits.

6.5 Carbon nanotube composites

6.5.1 Introduction

The potentially outstanding mechanical properties of carbon nanotubes discussed above will be of little value unless they can be incorporated into a

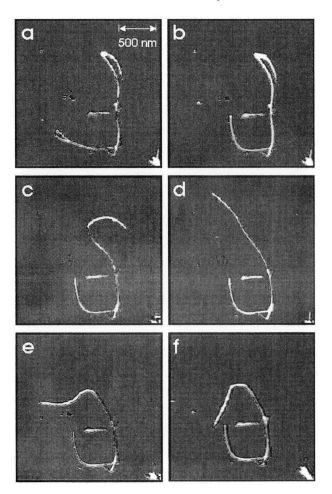

Fig. 6.10. Formation of the Greek letter Θ by manipulation of nanotubes on a silicon substrate using AFM (6.24).

matrix. Composite materials containing carbon fibres are, of course, already quite widely used in applications ranging from aerospace to sports equipment (6.1–6.4). In such materials the matrix can be plastic, epoxy, metal or carbon. The incorporation of carbon fibres into a matrix not only confers strength and elasticity to the material but also greatly enhances toughness, that is its ability to resist cracking.

The usual method of incorporating small, discontinuous fibres into a matrix is firstly to shape the fibres into a 'preform' and then to add the liquid matrix material under pressure. Adding the fibres to the liquid is not usually an option, since even quite small concentrations of fibres can have a thickening effect which makes mixing all but impossible (6.25). However, it should be noted that adding a matrix material to an array of nanoscale fibres may not be without difficulties. As a result of surface tension, the amount of pressure needed to produce infiltration into a fibre array increases as the spacing between the fibres decreases. Since carbon nanotubes are far smaller than normal reinforcing fibres, the spacing between the tubes will be far less than that for normal fibres, for a given volume fraction. Thus, the pressure required for infiltration may be very high. Having said this, the experimental work which has been carried out to date on the preparation of carbon nanotube composites indicates that achieving sufficient infiltration is not a serious problem.

In the following sections, some of the other factors relating to the production and properties of nanotube-reinforced composite materials are discussed. Firstly, the potential problems involved in achieving a bond between the tubes and the matrix are considered. The importance of the fibre aspect ratio is then discussed, and experimental work on the incorporation of carbon nanotubes into matrices reviewed. Finally, a brief discussion is given of the possible applications of carbon nanotube composites.

6.5.2 *Bonding between nanotubes and matrix*

It is generally believed that most multiwalled carbon nanotubes have a 'Russian doll' structure, in which each constituent tubule is only bonded to its neighbours by weak van der Waals forces (see Chapter 3). This immediately raises a problem when one is considering incorporating carbon nanotubes into matrices: how does one form a strong bond between the matrix and all the constituent tubules of a multiwalled nanotube? Bonding the outermost tube to the matrix would not anchor the inner tubes so that, in tension, the outer tube might simply slip away from the inner ones. It is not immediately obvious how this problem can be overcome, although the effects of intertube slipping might be less serious in a composite subject to bending or compression than to one in tension. Also, the presence of defects in the multiwalled structure might help in forming stronger intertube bonds. Two other questions which need to be addressed when considering the preparation of nanotube-reinforced composites are the wetting behaviour of nanotube surfaces, and the effect of chemical functionalisation on a tube's mechanical properties. We consider wetting behaviour first.

Experiments aimed at the opening and filling of nanotubes, described in the previous chapter, have provided important information on the wetting of nanotube surfaces. Thus, we know that metals like nickel wet the surfaces of nanotubes relatively poorly. As noted in Section 5.6.6, Dujardin, Ebbesen and

Fig. 6.11. Nanotubes treated with molten V_2O_5, showing partial filling and coating of tubes (6.27).

colleagues have shown that the 'cutoff' surface energy, γ, below which wetting of nanotubes would occur, lies somewhere between 100 and 200 mN/m (6.26). Thus, wetting of untreated nanotubes with a metal such as aluminium, which has a surface energy of around 840 mN/m should be very poor. On the other hand, work by Ajayan and colleagues has shown that nanotubes can be readily wetted by molten vanadium oxide (surface energy $\sim$ 80 mN/m) (6.27, 6.28). In this work, which was mentioned in the previous chapter (p. 172), a sample of opened nanotubes was mixed with V_2O_5 powder in a ratio of 1:1, and the mixture was heated to a temperature above the melting point of the oxide. It was found that the oxide entered the opened tubes, and formed thin layers on the tube surfaces as can be seen in Fig. 6.11. A similar effect has been observed by the Oxford group in experiments with molten molybdenum oxide (6.29). These studies suggest that nanotube–oxide composites could be prepared quite simply using oxides with suitable properties.

Where the surface characteristics of a matrix material are unfavourable, it may be possible to improve the wetting behaviour of the nanotubes by chemical treatment. This idea has been explored by several groups. For example, Malcolm Green's group in Oxford have shown that palladium ions interact strongly with nanotubes which have been treated with nitric acid (6.30). Similarly, Ebbesen and colleagues showed that deposition of uranium, yttrium and other metal particles on the surfaces of carbon nanotubes could be enhanced if the tubes were pre-oxidised (6.31).

Of course, it must be recognised that chemical functionalisation of nanotube surfaces will inevitably affect their mechanical properties. This problem has been discussed by Ajay Garg and Susan Sinnott of the University of Kentucky (6.32). These authors used classical molecular dynamics simulations to examine the effect of covalent chemical attachments on the stiffness of single-walled nanotubes. The maximum compressive (buckling) force for various functionalised and non-functionalised tubes was calculated. It was found that covalent chemical attachments decrease the maximum buckling force by about 15% regardless of tubule helical structure or radius. Presumably the effect in tension would be similar. Considering the exceptionally high Young's moduli of nanotubes, a reduction of about 15% in stiffness is not catastrophic and suggests that functionalisation could be used to anchor tubes in a matrix without a major loss of mechanical properties.

6.5.3 Aspect ratio

In order to make full use of the stiffness of the fibres in a composite, it must be possible for the fibre stress to reach its maximum value, σ_{max} (i.e. its breaking stress), through shear transfer (6.33). Thus, for maximum stress transfer to occur, the fibres must have a certain minimum aspect ratio. The minimum aspect ratio required for maximum stress transfer to occur is given by l_c/d, where l_c is the 'critical length' of the fibres and d is the diameter. Kelly (6.34) has shown that, for a unidirectional composite, the minimum fibre aspect ratio is given by

$$\frac{l_c}{d} = \frac{\sigma_{max}}{2\tau}$$

where τ is the interfacial shear stress. Thus, the stronger the fibres, the longer they need to be in order for the stress to reach its maximum value, for a given diameter. Fibres which are longer than l_c will be as effective as continuous fibres in stiffening the matrix.

We can use the above expression to obtain a very approximate figure for the critical length for carbon nanotubes in a composite material. First of all we need a figure for the breaking stress of nanotubes. No experimental measurements have yet been made, but simulations by the NCSU group (see above) suggest that a figure of around 100 GPa may be reasonable. If we now take the value for the interfacial shear stress of a typical polymer matrix to be ~ 50 MPa, and the typical diameter of a nanotube to be 10 nm, this gives a value for l_c of approximately 10 μm, i.e. an aspect ratio of 1000:1. This corresponds approximately to the maximum aspect ratio observed in a typical sample of nanotubes produced by arc-evaporation. Therefore, in order to exploit the full potential of nanotubes as components of polymer composites, methods for producing longer tubes may be needed. Higher aspect ratios would not be required, however, with other matrices such as ceramics, which have lower interfacial shear stresses.

6.5.4 Experiments on incorporating nanotubes into a matrix

To date, relatively little experimental work has been carried out on incorporating carbon nanotubes into matrices. In one of the earliest studies, Ajayan and colleagues embedded purified tubes into an epoxy resin (6.35), although their initial aim in doing so was not to produce a composite material. They were interested in obtaining cross-sectional images of nanotubes and attempted to achieve this by embedding purified tubes into an epoxy resin and then cutting the hardened 'composite' into thin slices with a diamond knife. However, no cross-sections were observed following this treatment; instead, the nanotubes were found to have become aligned in the direction of the knife movement, as shown in Fig. 6.12. According to Ajayan *et al.*, the alignment is primarily a consequence of extensional or shear flow of the matrix produced by the cutting, although in some cases tubes appear to come into direct contact with the knife and are pulled out of the matrix and oriented unidirectionally on the surface. They likened the effect to the molecular alignment of liquid crystals which can be achieved by rubbing the crystals between flat surfaces.

The main importance of this work is in providing a graphic demonstration that unidirectional carbon nanotube composites can be prepared. In a composite material it is often desirable for the fibres to be oriented in a specific way. This is relatively straightforward for conventional fibres, which are generally produced as continuous tows and are large enough to be aligned or woven mechanically. For carbon nanotubes, with lengths in the micrometre range and diameters of a few tens of nanometres or less, these techniques are clearly not applicable. The work of Ajayan *et al.* shows that alignment of carbon nanotubes can be achieved during processing of the composite.

The synthesis of carbon nanotube composites has also been studied by Toru Kuzumaki and colleagues of the University of Tokyo (6.36, 6.37). This group have succeeded in preparing carbon nanotube/aluminium composites (6.36), despite the high surface energy of aluminium, noted above. Their method involved mixing a nanotube sample with a fine aluminium powder, mounting the mixture in a 6 mm silver sheath and then drawing and heating the wire at 700 °C in a vacuum furnace. The result was a composite wire in which the nanotubes were partially aligned along the axial direction. This method of

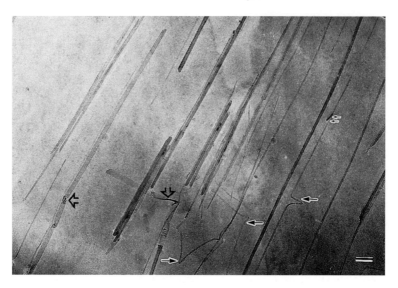

Fig. 6.12. Alignment of nanotubes in a polymer matrix following cutting with micro-tome (6.35).

aligning fibres using extensional flow is well established in composite prepara-tion (6.33). The Tokyo group have also prepared nanotube/C_{60} composites using a similar method (6.37).

A group from Ireland and the US have described the synthesis of a compos-ite material containing arc-produced multiwalled nanotubes in a luminescent polymer, PmPV (6.38). Good wetting of the tubes by the polymer was achieved, and the composite was found to have useful properties, as described in the next section.

6.5.5 *Applications of nanotube-containing composites*

The most obvious applications of carbon nanotube composites would be as structural materials, but the extremely high production costs mean that such applications are out of the question at present. At the time of writing, the market price of an unpurified sample of carbon nanotubes, produced by the arc-evaporation method, is of the order of $20 per gram. This compares with somewhere around a few dollars per kilogram for carbon fibres, and it should be remembered that carbon fibres themselves have proved too expensive for mass-market applications. Evidently, much improved production methods (perhaps based on some of the new techniques described in Chapter 2) will be needed before carbon nanotube-based skyscrapers and aircraft become a reality. However, there are other, more specialised, applications for nanotube-containing composites, and here the high cost would be less of a problem. One area which is attracting interest involves incorporating nanotubes in matrices to improve their electrical conductivity. Currently, both carbon black and carbon fibres are used as conductive fillers in polymer matrices. Work is undoubtedly being carried out in several laboratories on nanotube-containing composites for electrical applications, but little has yet appeared in the open literature.

As mentioned above, the properties of a composite material containing nanotubes in a luminescent polymer have been investigated (6.38). In such systems, it is often desirable to increase electrical conductivity. However, previous attempts to achieve this by doping have resulted in degradation of the optical properties of the polymer. In the case of nanotubes, an increase in the electrical conductivity of the polymer by up to eight orders of magnitude was seen, with no concomitant degradation of optical properties. This was apparently because the tubes acted as nanometric heat sinks, preventing the build up of large thermal effects, which damage these conjugated systems. It was also shown in this work that the composite can be used as the emissive layer in an organic light-emitting diode, which promises future applications in optoelectronics.

Incorporating carbon nanotubes into matrices other than polymers may also be beneficial. For example, work by Ebbesen and colleagues has shown that nanotubes can be embedded in a superconducting oxide (Bi2212), resulting in superior superconducting properties (6.39). This was attributed to strong pinning of vortex lines along the length of the tubes.

6.6 Nanotubes as tips for scanning probe microscopes

The outstanding mechanical properties and unique geometry of carbon nanotubes have led to suggestions that they would make ideal candidates as tips for scanning probe microscopes (SPMs). Certainly nanotubes would seem to have advantages over the tips currently used for atomic force microscopy (AFM), which typically consist of microfabricated pyramids of silicon or silicon nitride. These tips can be relatively 'blunt' on the scale of the features which are being imaged, and are thus often unable to probe narrow crevices on the specimen surface. The elongated shape and small diameter of carbon nanotubes should enable them to probe the most narrow of fissures. There is also the possibility of functionalising the nanotubes in order to carry out 'chemical force microscopy'. There are potential problems with using nanotubes for atomic resolution imaging, however, due to the relatively large thermal vibrations observed at room temperature (see Section 6.4.3 above).

Smalley and co-workers were the first to demonstrate the use of nanotubes as SPM tips. In a paper published in late 1996 (6.40), they described a method of attaching bundles of multiwalled nanotubes to the tips of commercial silicon pyramids, and then drawing out a single tube from the bundle to act as the imaging probe. The nanotube tips were then used to obtain tapping-mode AFM images of a patterned film on a silicon wafer. It was found that the tips could reach to the bottom of deep trenches in the film, and thus produce much more realistic images than those obtained using pyramidal tips. Because the nanotube tips were electrically conducting, they could be used for scanning tunnelling microscopy (STM) as well as AFM, and the Smalley group obtained atomic resolution images of TaS_2 surfaces using the tips in STM mode.

Charles Lieber and colleagues from Harvard have investigated the possibility of using carbon nanotube tips for imaging and analysing biological systems (6.41–6.43). Their method of fixing bundles of nanotubes to silicon pyramids was similar to that used by Smalley and co-workers, and both multiwalled and single-walled tubes were used. In the first study (6.41), the nanotube tips were used to obtain tapping-mode AFM images of amyloid fibrils. The resulting images appeared to show greater detail than those obtained using conventional tips, and the nanotube probes proved to be both robust and relatively resistant to contamination.

In further work (6.42, 6.43), Lieber's group prepared functionalised nanotube tips, and then used these to sense specific interactions with functional groups on substrates. The tubes were initially oxidised, removing the caps and resulting in the formation of carboxyl surface groups. The carboxyl-terminated tubes were then used to carry out chemically sensitive imaging of surfaces patterned with different molecules. Tubes terminated with amine groups were used in a similar way. In addition to chemical imaging, Lieber and co-workers used nanotube tips to investigate interactions between biological molecules, specifically the ligand–receptor interaction of biotin with streptavidin. Biotin ligand was covalently linked to nanotube tips by the formation of amide bonds, and the modified tips were then used to probe immobilised streptavidin molecules adsorbed on mica. In this way it proved possible to measure the binding forces between the biotin–streptavidin pairs.

Nanotubes also have potential for nanolithography. Hongjie Dai and colleagues have used nanotube AFM tips to draw patterns and words on silicon oxide surfaces, as shown in Fig. 6.13 (6.44). The nanotubes were found to have greatly improved wear properties compared with conventional tips. Graphite has been used as a writing material since time immemorial. Perhaps graphitic nanotubes will prove to be the ideal writing implements for the nanoworld.

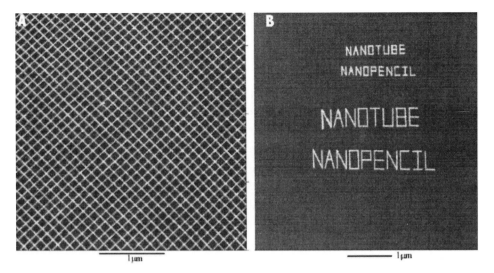

Fig. 6.13. Examples of nanolithography produced using a nanotube AFM tip (6.44). (a) Pattern of lines with 100 nm spacing, (b) some nanowriting.

6.7 Discussion

It is now clearly established that carbon nanotubes possess quite unique mechanical properties. The TEM work of Treacy, Ebbesen and Gibson, and subsequent AFM studies by the Lieber group and others have produced estimates for the Young's modulus that are equal to, or greater than, the accepted value for a graphene sheet of 1.06 TPa. Thus, carbon nanotubes are much the stiffest of all known materials. Skilful AFM studies have also enabled the behaviour of nanotubes under high stress to be probed. These have shown that the flexibility of the graphene cylinders enables nanotubes to accommodate extreme deformations without fracturing, and that in many cases they can recover apparently unscathed from such deformations. This sets them apart from conventional carbon fibres, and all other currently known fibres, which are much more susceptible to fracture when bent or twisted.

These remarkable mechanical characteristics will undoubtedly lead to many important applications. Indeed, nanotubes are already being used with great success as tips for scanning probe microscopes, not only for topological imaging but also for chemical imaging and to investigate interactions between biological molecules. Very recent work by de Heer and colleagues has pointed to other ways in which the stiffness of nanotubes might enable them to be used to probe the nanoworld (6.45). These workers applied alternating voltages to nanotubes inside an electron microscope, causing them to vibrate. By adjusting the frequency of the applied potential they were able to excite the nanotubes resonantly at their fundamental frequency and at higher harmonics. In this way they could determine not only the tubes' moduli, but could also measure the masses of carbon nanoparticles which were attached to the tubes. They suggest that this 'nanobalance' approach could be applied to other particles of similar dimensions such as viruses.

Carbon nanotubes should also be excellent candidates for incorporation into composite materials, and initial studies here have been very promising. It should be recognised, however, that the incorporation of molecular scale fibres like carbon nanotubes in composites is a new field, and there are still major obstacles to overcome. In this connection, there might be lessons to learn from nature. After all, many structural biomaterials, like bone, antler and spider silk can be thought of as highly sophisticated nanocomposites. Perhaps the availability of carbon nanotubes and other nanofibres will enable us to construct materials as exceptional as some of these.

References

(6.1) E. Fitzer (ed.), *Carbon fibres and their composites*, Springer-Verlag, Berlin, 1986.
(6.2) M. S. Dresselhaus, G. Dresselhaus, K. Sugihara, I. L. Spain and H. A. Goldberg, *Graphite fibers and filaments*, Springer-Verlag, Berlin, 1988.
(6.3) D. J. Johnson in *Introduction to carbon science*, ed. H. Marsh, Butterworths, London, 1989, p. 197.
(6.4) G. Savage, *Carbon–carbon composites*, Chapman and Hall, London, 1992.
(6.5) G. Overney, W. Zhong and D. Tomanek, 'Structural rigidity and low frequency vibrational modes of long carbon tubules', *Z. Phys. D*, **27**, 93 (1993).
(6.6) M. M. J. Treacy, T. W. Ebbesen and J. M. Gibson, 'Exceptionally high Young's modulus observed for individual carbon nanotubes', *Nature*, **381**, 678 (1996).
(6.7) R. W. Clark, *Edison*, Macdonald and James, London, 1977.
(6.8) P. J. F. Harris, 'Structure of non-graphitising carbons', *International Materials Reviews*, **42**, 206 (1997).
(6.9) R. Bacon, 'Growth, structure and properties of graphite whiskers', *J. Appl. Phys.*, **31**, 283 (1960).
(6.10) M. Endo, 'Grow carbon fibres in the vapor phase', *Chemtech*, **18**, 568 (1988) (September issue).
(6.11) J. P. Lu, 'Elastic properties of carbon nanotubes and nanoropes', *Phys. Rev. Lett.*, **79**, 1297 (1997).
(6.12) E. Hernández, C. Goze, P. Bernier, A. Rubio, 'Elastic properties of C and $B_xC_yN_z$ composite nanotubes', *Phys. Rev. Lett.*, **80**, 4502 (1998).
(6.13) B. I. Yakobson, C. J. Brabec and J. Bernholc, 'Nanomechanics of carbon tubes: instabilities beyond linear response', *Phys. Rev. Lett.*, **76**, 2511 (1996).
(6.14) B. I. Yakobson, M. P. Campbell, C. J. Brabec and J. Bernholc, 'High strain rate fracture and C-chain unraveling in carbon nanotubes', *Comp. Mat. Sci.*, **8**, 341 (1997).

(6.15) J. F. Despres, E. Daguerre and K. Lafdi, 'Flexibility of graphene layers in carbon nanotubes', *Carbon*, **33**, 87 (1995).

(6.16) R. S. Ruoff and D. C. Lorents, 'Mechanical and thermal properties of carbon nanotubes', *Carbon*, **33**, 925 (1995).

(6.17) K. L. Lu, R. M. Lago, Y. K. Chen, M. L. H. Green, P. J. F. Harris and S. C. Tsang, 'Mechanical damage of carbon nanotubes by ultrasound', *Carbon*, **34**, 814 (1996).

(6.18) S. Iijima, C. Brabec, A. Maiti and J. Bernholc, 'Structural flexibility of carbon nanotubes', *J. Chem. Phys.*, **104**, 2089 (1996).

(6.19) B. I. Yakobson and R. E. Smalley, 'Fullerene nanotubes: $C_{1,000,000}$ and beyond', *American Scientist*, **85**, 324 (1997).

(6.20) T. Kuzumaki, T. Hayashi, H. Ichinose, K. Miyazawa, K. Ito and Y. Ishida, '*In-situ* observed deformation of carbon nanotubes', *Philos. Mag. A*, **77**, 1461 (1998).

(6.21) O. Lourie, D. M. Cox and H. D. Wagner, 'Buckling and collapse of embedded carbon nanotubes', *Phys. Rev. Lett.*, **81**, 1638 (1998).

(6.22) E. W. Wong, P. E. Sheehan and C. M. Lieber, 'Nanobeam mechanics: elasticity, strength, and toughness of nanorods and nanotubes', *Science*, **277**, 1971 (1997).

(6.23) M. R. Falvo, G. J. Clary, R. M. Taylor, V. Chi, F. P. Brooks, S. Washburn and R. Superfine, 'Bending and buckling of carbon nanotubes under large strain', *Nature*, **389**, 582 (1997).

(6.24) T. Hertel, R. Martel and P. Avouris, 'Manipulation of individual carbon nanotubes and their interaction with surfaces', *J. Phys. Chem. B*, **102**, 910 (1998).

(6.25) J. E. Gordon, *The new science of strong materials*, Penguin, Harmondsworth, 1968.

(6.26) E. Dujardin, T. W. Ebbesen, H. Hiura and K. Tanigaki, 'Capillarity and wetting of carbon nanotubes', *Science*, **265**, 1850 (1994).

(6.27) P. M. Ajayan, O. Stephan, Ph. Redlich and C. Colliex 'Carbon nanotubes as removable templates for metal oxide nanocomposites and nanostructures', *Nature*, **375**, 564 (1995).

(6.28) P. M. Ajayan, P. Redlich and M. Rühle, 'Structure of carbon nanotube-based composites', *J. Microscopy*, **185**, 275 (1997).

(6.29) Y. K. Chen, M. L. H. Green and S. C. Tsang, 'Synthesis of carbon nanotubes filled with long continuous crystals of molybdenum oxides', *J. Chem. Soc., Chem. Commun.*, 2489 (1996).

(6.30) R. M. Lago, S. C. Tsang, K. Lu, Y. K. Chen and M. L. H. Green, 'Filling carbon nanotubes with small palladium metal crystallites – the effect of surface acid groups', *J. Chem. Soc., Chem. Commun.*, 1355 (1995).

(6.31) T. W. Ebbesen, H. Hiura, M. E. Bisher, M. M. J. Treacy, J. L. Shreeve-Keyer and R. C. Haushalter, 'Decoration of carbon nanotubes', *Advanced Materials*, **8**, 155 (1996).

(6.32) A. Garg and S. B. Sinnott, 'Effect of chemical functionalization on the mechanical properties of carbon nanotubes', *Chem. Phys. Lett.*, **295**, 273 (1998).

(6.33) D. Hull, *An introduction to composite materials*, Cambridge University Press, Cambridge, 1981.

(6.34) A. Kelly, 'Interface effects and the work of fracture of a fibrous composite', *Proc. R. Soc. Lond. A*, **319**, 95 (1970).

(6.35) P. M. Ajayan, O. Stephan, C. Colliex and D. Trauth, 'Aligned carbon nanotube arrays formed by cutting a polymer resin–nanotube composite', *Science*, **265**, 1212 (1994).

(6.36) T. Kuzumaki, K. Miyazawa, H. Ichinose and K. Ito, 'Processing of carbon nanotube reinforced aluminum composite', *J. Mater. Res.*, **13**, 2445 (1998).

(6.37) T. Kuzumaki, T. Hayashi, K. Miyazawa, H. Ichinose, K. Ito and Y. Ishida, 'Processing of ductile carbon nanotube/C_{60} composite', *Materials Trans., JIM*, **39**, 574 (1998).

(6.38) S. A. Curran, P. M. Ajayan, W. J. Blau, D. L. Carroll, J. N. Coleman, A. B. Dalton, A. P. Davey, A. Drury, B. McCarthy, S. Maier and A. A. Strevens, 'A composite from poly(m-phenylenevinylene-co-2,5-dioctoxy- p-phenylenevinylene) and carbon nanotubes: A novel material for molecular optoelectronics', *Advanced Materials*, **10**, 1091 (1998).

(6.39) K. Fossheim, E. D. Tuset, T. W. Ebbesen, M. M. J. Treacy, and J. Schwartz, 'Enhanced flux pinning in $Bi_2Sr_2CaCu_2O_{8+x}$ superconductor with embedded carbon nanotubes', *Physica C*, **248**, 195 (1995).

(6.40) H. Dai, J. H. Hafner, A. G. Rinzler, D. T. Colbert and R. E. Smalley, 'Nanotubes as nanoprobes in scanning probe microscopy', *Nature*, **384**, 147 (1996).

(6.41) S. S. Wong, J. D. Harper, P. T. Lansbury and C. M. Lieber, 'Carbon nanotube tips: high-resolution probes for imaging biological systems', *J. Amer. Chem. Soc.*, **120**, 603 (1998).

(6.42) S. S. Wong, E. Joselevich, A. T. Woolley, C. L. Cheung and C. M. Lieber, 'Covalently functionalized nanotubes as nanometre-sized probes in chemistry and biology', *Nature*, **394**, 52 (1998).

(6.43) S. S. Wong, A. T. Woolley, E. Joselevich, C. L. Cheung and C. M. Lieber, 'Covalently-functionalized single-walled carbon nanotube probe tips for chemical force microscopy', *J. Amer. Chem. Soc.*, **120**, 8557 (1998).

(6.44) H. Dai, N. Franklin, and J. Han, 'Exploiting the properties of carbon nanotubes for nanolithography', *Appl. Phys. Lett.*, **73**, 1508 (1998).

(6.45) P. Poncharal, Z. L. Wang, D. Ugarte and W. A. de Heer, 'Electrostatic deflections and electromechanical resonances of carbon nanotubes', *Science*, **283**, 1513 (1999).

7

Curved crystals, inorganic fullerenes and nanorods

Daedalus, who published detailed predictions of hollow graphite molecules years ago, points out that other substances with sheet molecules, such as molybdenum disulphide and mica, should also fragment into flakes and curl up under the same treatment.

David Jones, *New Scientist*, 24 April 1986

Following the discovery of C_{60} in 1985, the ever-inventive Daedalus sought to maintain his lead over the experimentalists by putting forward the idea of *inorganic* fullerenes. A number of inorganic compounds form graphite-like layered crystals, so the idea of closed structures based on these materials certainly seemed plausible. In fact curved inorganic crystals had been known since the 1950s, having been proposed theoretically by Linus Pauling in 1930 (7.1). The most striking example occurs in the chrysotile form of serpentine, the primary constituent of most asbestos, which occurs primarily as tightly curled tubular structures. However, such structures differ from fullerenes in that the curvature derives solely from a structural mismatch between adjacent layers. In a true inorganic fullerene the curvature would be associated with point defects equivalent to the pentagons in carbon fullerenes. Particles of tungsten disulphide and other dichalcogenides with just this kind of structure have now been synthesised (7.2). Yet again the discovery was serendipitous, and occurred as a byproduct of attempts by Reshef Tenne and co-workers at the Weizmann Institute in Israel to prepare thin films of tungsten disulphide for use in solar cells. Subsequent work has shown that the inorganic fullerenes have exceptional lubricating properties.

Boron nitride is another material which exists in a graphite-like layered form, and in the 1980s various groups showed that graphite hybrids containing C, B and N could be prepared (e.g. 7.3). Following the discovery of carbon nanotubes, theoreticians predicted that BN and BCN nanotubes should be stable (7.4, 7.5). This was confirmed a short time later when both types of nanotube were successfully produced using variations of the Krät-schmer–Huffman technique (7.6, 7.7). At about the same time a 'catalytic' synthesis of BN nanotubes was also demonstrated (7.8), and subsequently various other methods of producing BN and BCN nanostructures have been developed. A number of authors have discussed the properties of these new structures, which should be quite different from those of their pure carbon analogues.

The use of carbon nanotubes as templates for the synthesis of inorganic nanostructures was first explored by Pulickel Ajayan and colleagues in 1995 (7.9). These workers showed that nanotubes could be coated with thin uniform layers of vanadium pentoxide by treatment with the molten oxide, as discussed in the previous chapter (Section 6.5.2). They also showed that the underlying tubes could be removed by oxidation at a temperature below the melting point of V_2O_5, leaving free-standing oxide tubules. A short time later, a different approach to the synthesis of inorganic nanofibres was described by a group from Harvard led by Charles Lieber (7.10). These workers described a method for converting carbon nanotubes into carbide 'nanorods' by reacting them with volatile oxide and/or halide species. This technique might lead to the synthesis of carbide nanostructures with useful magnetic, electronic or structural applications.

The main part of this chapter is concerned with the synthesis, structure and properties of the new fullerene-like inorganic nanostructures. Firstly, though, a brief description is given of two naturally occurring inorganic materials, chrysotile asbestos and imogolite, which have striking similarities to carbon nanotubes.

7.1 Chrysotile and imogolite

In his 1930 paper Pauling pointed out that if the two faces of a constituent layer in a layered mineral are not equivalent, the structural mismatch between adjacent layers could lead to strain, and therefore bending. This was confirmed in the 1950s, when it was recognised that unusual features in X-ray diffraction patterns of the serpentine group of magnesium silicate minerals could only be interpreted by assuming curvature of the crystal planes (7.11). Since that time, the application of high resolution electron microscopy to such minerals has greatly enhanced our understanding of their structure (7.12–7.15).

The serpentine minerals make up one of the two groups commonly referred to as asbestos (the other being amphibole). They have been intensively researched, mainly because of their carcinogenic properties. Figure 7.1 shows

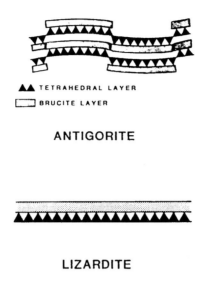

ANTIGORITE

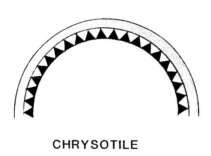

LIZARDITE

CHRYSOTILE

Fig. 7.1. Structures of the minerals which make up the serpentine group.

schematically the structures of the minerals which make up the serpentine group. The black triangles represent cations (normally Si) which are coordinated to four oxygens, while the shaded layer contains cations (normally Mg) coordinated to six oxygens. In the case of the antigorite structure, there is a small misfit between the tetrahedral silicate sheet and the octahedral hydroxide sheet, producing a corrugated structure, while in lizardite there is no structural mismatch, and the layers are planar. The most pronounced effect of the structural mismatch occurs in chrysotile (or 'white asbestos') where continuous curvature is observed.

A high resolution electron micrograph, by David Veblen and Peter Buseck (7.14), showing a region of chrysotile intergrown with planar varieties of serpentine, is shown in Fig. 7.2. In its pure form, chrysotile can grow into long tubules which are strikingly similar in appearance to carbon nanotubes, as

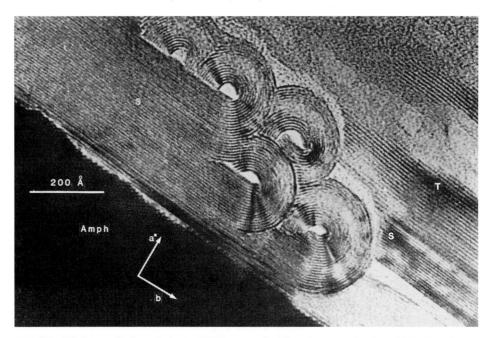

Fig. 7.2. High resolution electron microscope image of serpentine in which the planar form (lizardite) is terminated with rolls of chrysotile (7.14).

shown in Fig. 7.3, taken from the work of John Hutchison (7.12). However, there are important differences between these structures and carbon nanotubes. Firstly, images of chrysotile tubules in cross-section show that although they sometimes have a concentric structure, they much more commonly consist of single or multiple scrolls, as shown in Fig. 7.4. As discussed in Chapter 3, all the evidence leads us to believe that nanotubes have a concentric structure. Secondly, capped chrysotile tubes are never observed. This emphasises the difference between the two types of tubular crystal. In the case of carbon nanotubes, the caps appear to play an essential role in forming the tubular structure, whereas the curvature of chrysotile tubes derives solely from their intrinsic crystal structure.

If chrysotile asbestos fibres closely resemble multiwalled carbon nanotubes, at least in appearance, then imogolite has remarkable similarities to single-walled nanotubes. Imogolite is a clay mineral which is mostly found in soils derived from volcanic ash, and in weathered pumices. Analysis by HREM and other techniques has established that it is made up of aluminosilicate tubes approximately 2 nm in diameter (7.16, 7.17). A micrograph of this material is shown in Fig. 7.5 and the similarities with SWNTs are clear. The tubes are uniform in diameter and, as a result, have a tendency to form aligned bundles,

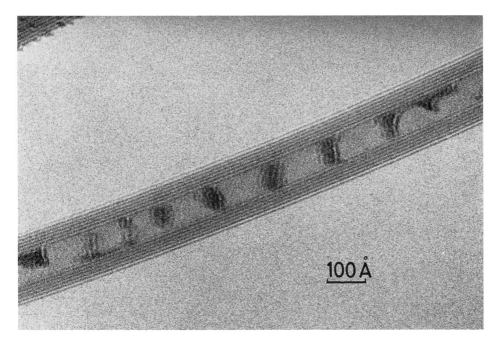

Fig. 7.3. Typical chrysotile fibril, containing blobs of amorphous silica (7.12).

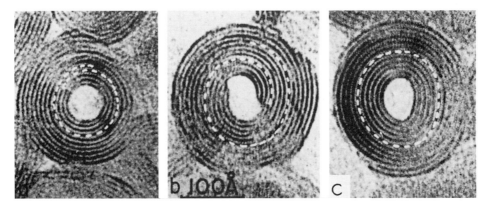

Fig. 7.4. Cross-sectional images of chrysotile fibrils showing various multilayer structures. (a) Single layer spiral, (b) multiple spiral, (c) concentric layers (7.12).

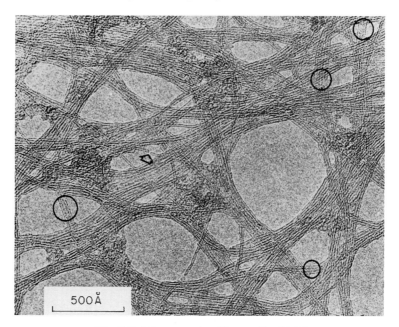

Fig. 7.5. Micrograph of imogolite (7.16).

just like the 'ropes' of SWNTs discussed in Chapters 2 and 3. As can be seen, they also have the ability to bend without fracture. Further work on these fascinating inorganic SWNTs would be of value.

7.2 Inorganic fullerenes from layered metal dichalcogenides

7.2.1 Synthesis of chalcogenide fullerenes

As noted above, Tenne and co-workers first produced inorganic fullerenes while attempting to prepare thin films of tungsten disulphide. Their method involved annealing a thin film of tungsten at 1000 °C in an atmosphere of hydrogen sulphide. When examining the resulting structures by transmission electron microscopy, they noticed some unusual tungsten disulphide particles, but did not pay them any great attention. Only when they saw Iijima's images of nested carbon nanotubes in late 1991 were they prompted to take a closer look at the crystallites. Although most of the WS_2 particles were approximately equiaxed in shape rather than tubular, high resolution electron microscopy showed their structures to be strikingly similar to their all-carbon analogues, and the work was immediately written up for *Nature* (7.2).

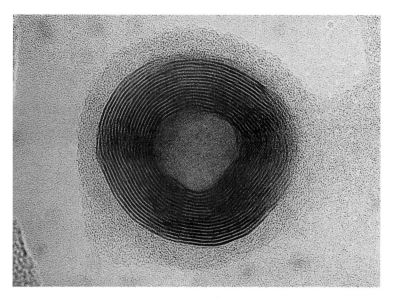

Fig. 7.6. Typical tungsten disulphide nanoparticle, courtesy Reshef Tenne. The spacing between the layers is approximately 0.62 nm.

Examples of the fullerene-like tungsten disulphide particles are shown in Figs. 7.6 and 7.7. The particles ranged in size from approximately 10 nm to 100 nm, and had a variety of shapes. In some cases the particles were clearly tubular, with capped ends like those observed in carbon nanotubes, as shown in Fig. 7.7. Subsequently, nanoparticles of MoS_2, $MoSe_2$ and WS_2 were prepared (7.18–7.22). In the case of MoS_2, it was found that a much better yield of nanoparticles could be achieved by a gas-phase reaction between molybdenum suboxide (MoO_{3-x}) and H_2S than by gas–solid reaction (7.20). Such a reaction can be carried out at relatively low temperatures since MoO_{3-x} sublimes at around 650 °C. However, rather careful control of the reaction conditions is required to produce the sub-stoichiometric oxide MoO_{3-x} from the MoO_3 starting material. It was also found that the nature of the product depended sensitively on the flow conditions in the reactor, with turbulent flow conditions producing the highest yield of inorganic fullerenes, apparently due to better mixing of reactants. High yields of nanotubes could also be produced under certain conditions.

Tenne and colleagues believe that amorphous chalcogenides are the precursors for inorganic fullerene formation, and have supported this by showing that $a\text{-}WS_3$ can be converted into nanoparticles by controlled heating at ~ 850 °C. They also found that a sample of $a\text{-}WS_3$ which had been simply stored in a drawer for over two years contained nanoparticles. In the light of

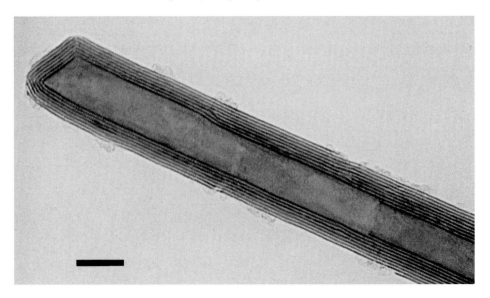

Fig. 7.7. Capped tungsten disulphide nanotube (7.2). Scale bar 10 nm.

these observations it seems very surprising that inorganic fullerenes were not observed many years ago.

7.2.2 Structure of chalcogenide fullerenes

Tungsten disulphide has a graphite-like layered structure consisting of alternating hexagonal planes of tungstens and sulphurs, with an interlayer distance of 0.62 nm, as shown in Fig. 7.8. Unlike chrysotile asbestos, and some of the other layered materials mentioned above, there is no propensity for the layers to curve. This indicates that the curvature arises from defects in the hexagonal planes, and is consistent with the observation that the WS_2 nanoparticles and tubes are almost invariably closed. Tenne and colleagues believe that these defects are probably triangular or rhombic in shape, rather than pentagonal as in fullerenes, and that they could arise from a single tungsten vacancy (7.18). Individual triangular and rhombic defects are illustrated in Fig. 7.9. In order to determine the overall shape of particles containing these defects, we can invoke Euler's theory. This states that a closed polyhedron will require four triangular faces or six quadrilateral faces. A particle with four triangular point defects would be approximately tetrahedral in shape, while a particle with six quadrilateral point defects would have a shape approximating to a flattened octahedron.

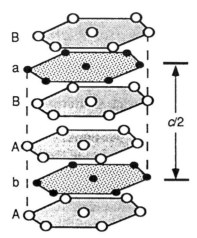

Fig. 7.8. The structure of tungsten disulphide (7.2). Open circles, sulphur; closed circles, tungsten.

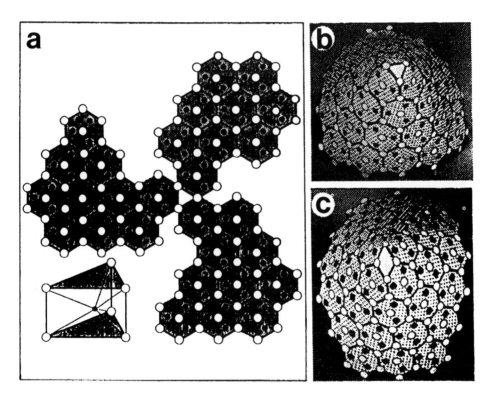

Fig. 7.9. (a) Drawing of W–S–W layer, and view of trigonal prism of 2H-WS$_2$ lattice. Open circles, sulphur; closed circles, tungsten. (b) Drawing of conical apex formed by folding and joining three hexagons around a triangle. (c) Conical asymmetric apex formed by folding and joining four hexagons around a rhombus (7.18).

7.2.3 *Inorganic fullerenes as solid-state lubricants*

Although most lubricants are liquids, there are some situations (such as very high or low temperatures, or high vacuum) where the use of solid lubricants is required. Both graphite and transition metal dichalcogenides are good solid lubricants, since the weakness of their interplanar bonds allows shearing between layers. However, neither of these can be used in all types of environment. In the case of the dichalcogenides, the presence of 'dangling bonds' can limit their effectiveness in reactive environments. These dangling bonds occur at the edges of planes and at defects in the crystal structure, and can react rapidly with water or oxygen, resulting in greatly impaired lubricating properties. Graphite, on the other hand, requires a small amount of adsorbed material to facilitate the sliding of the layers, and loses its lubricating properties in a vacuum. There is therefore a need for solid lubricants capable of operating under a wider range of conditions.

Tenne and colleagues realised that metal dichalcogenide nanoparticles, with their spheroidal shapes, might be useful lubricants and in mid-1997 reported a series of experiments to test their tribological properties (7.22). Tungsten disulphide nanoparticles were prepared by the solid–gas reaction method and their lubricating properties compared with those of WS_2 and MoS_2 powders. The nanoparticles proved to be superior in all respects (friction, wear and lifetime of the lubricant) to the conventional powder samples. This indicates that the mechanism of lubrication for the WS_2 nanoparticles is quite different from that of the crystalline powders, and involves rolling of the particles rather than interplanar sliding. Tenne *et al.* noted that the hollow structure of the nanoparticles was also conducive to lubrication, since this enables the particles to deform elastically rather than inelastically, reducing the energy dissipation associated with friction. The closed nature of the nanoparticles, with few if any dangling bonds also rendered them less susceptible to chemical attack. Overall, the WS_2 nanoparticles appear to have considerable potential as lubricating materials.

In the early days of fullerene research, the possibility that C_{60} molecules might make ideal solid lubricants was frequently discussed. Unfortunately these ideas came to nothing when crystalline C_{60} turned out to be about as slippery as scouring powder! Following the work of Tenne and colleagues, maybe inorganic rather than carbon fullerenes will find an application as 'molecular ball bearings'.

7.3 Nanotubes and nanoparticles containing boron and nitrogen

7.3.1 Boron-carbon-nitride tubes

In 1994, workers from the Université de Paris-Sud and the Université de Montpellier in France reported the synthesis of nanoscale B, C and N-containing tubes (7.6). Their method involved placing a mixture of boron and graphite powder into a hollowed-out graphite anode and carrying out the arc-evaporation in an atmosphere of nitrogen. As is the case for pure carbon nanotubes, the BCN tubes were found inside the deposit which formed on the cathode. They were accompanied by BN and BCN sheets, glassy carbon and pure carbon nanotubes and nanoparticles; elemental analysis of these structures was carried out using EELS. Two types of B and N-containing nanotubes were present: relatively large diameter fibres (100–500 nm) with irregular thickening, and narrower tubes up to 100 μm in length. In both cases, the caps of the tubes were found to be poorly formed in comparison with the caps of pure carbon nanotubes, presumably because five-membered rings are less easy to form in BCN networks than in pure carbon networks. The difficulty in forming caps also probably explains why in some cases the BCN tubes grew to much greater lengths than usually observed for carbon nanotubes. A short time after the French work appeared, Marvin Cohen's group from the University of California at Berkeley also described the synthesis of $B_xC_yN_z$ nanotubes, this time using a graphite anode which had been drilled out to contain a BN rod, and arc-evaporating in helium (7.23).

Arc-evaporation is not the only method of producing BCN nanostructures; a pyrolytic synthesis method has been described by Mauricio Terrones and colleagues (7.24). This involved pyrolysis of $CH_3CN.BCl_3$ at 900–1000 °C over cobalt powder, and resulted in the formation of graphitic $B_xC_yN_z$ nanofibres and nanotubes possessing a range of morphologies. The cobalt particles appeared to play an important role in the growth process, possibly as catalysts.

In the BCN structures discussed so far, it has been assumed that the B and N atoms are incorporated into the graphite lattice. In 1997, French workers described the synthesis of nanotubes and nanoparticles which were apparently made up of all-carbon and all-BN layers, rather than layers containing all three elements (7.25). These structures were prepared by arc-evaporating a hafnium diboride rod with graphite in a nitrogen atmosphere. Elemental analysis showed that most of the tubes had a sandwich structure with carbon layers both in the centre and at the periphery, separated by a few BN layers. The reason for this unusual growth configuration is not clear.

7.3.2 Pure boron nitride tubes and nanoparticles

The synthesis of *pure* boron nitride filaments has now been reported by a number of groups. Structures ranging in diameter from several micrometres to a few nanometres have been produced. Workers from Ohio State University reported in 1993 that tubular BN structures with diameters up to 3 μm and lengths up to 100 μm could be prepared by heating amorphous BN to 1100 °C (7.26). These large tubes bear no resemblance to carbon nanotubes, but other workers soon showed that nanoscale BN tubes can also be formed. In 1994, French workers reported the synthesis of much smaller tubular boron nitride filaments using a catalytic method (7.8). This involved passing N_2 or NH_3 over boride catalysts such as ZrB_2 at a temperature of ~ 1100 °C. The filaments produced in this way were rather similar to catalytically produced carbon nanotubes, although somewhat larger in diameter (typically 100 nm). Transmission electron microscopy showed that the tubes were often segmented in structure rather than continuously tubular, and EELS analysis confirmed that the tubes consisted of boron nitride. The authors suggested that the mechanism of tube formation involved a reaction between N_2 or NH_3 and a boride particle to produce BN which is then extruded from the particle as a tube.

The synthesis of boron nitride nanotubes using an arc-evaporation technique was reported by the Berkeley group in 1995 (7.7). In order to avoid any possibility of contamination with carbon, they employed no graphite components in the synthesis. Thus, the anode consisted of a hollow rod of tungsten into which was inserted a pressed rod of hexagonal boron nitride, and the cathode was a cooled copper electrode. Evaporation was carried out under helium using conditions quite similar to those employed for carbon nanotube synthesis. Again, the nanotubes were found inside the cathodic deposit, where they were accompanied by deposits of tungsten and other ill-defined structures. Detailed examination of the nanotubes using HREM showed that they were multiwalled structures with inner diameters ranging from ~ 1–3 nm and outer diameters of ~ 6–8 nm, with lengths typically exceeding 200 nm. The interlayer distance was ~ 0.33 nm, consistent with the interplanar distance in bulk h-BN of 0.333 nm. None of the BN nanotubes were found to be capped. Instead, dense particles of tungsten or a tungsten-containing material were invariably found at the tips of the BN tubes, as shown in Fig. 7.10. The role of the tungsten-containing particles in the formation of BN nanotubes is not clear, but does not appear to be catalytic. The Berkeley group believe that the first stage of tube growth involves the particles becoming coated with BN layers, and that in some cases the layers become opened, and tube growth then

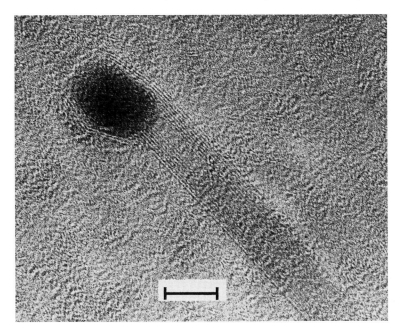

Fig. 7.10. Boron nitride nanotube with particle at tip (7.7). Scale bar 6 nm.

proceeds by addition of BN in the axial direction. This is supported by the observation that many BN-coated particles are observed in the arc-evaporated material. Termination may occur when the growing end of a tube accidentally impinges on another tungsten-containing particle.

Other workers have described the synthesis of BN nanotubes and nanoparticles using arc-evaporation. Terrones and colleagues described the structures produced by arcing h-BN and tantalum in a nitrogen atmosphere (7.27). Unlike the Berkeley group they found that in some cases the tubes produced in this way were capped. Annick Loiseau and colleagues have described HREM studies of BN tubes produced by arc-discharge between HfB_2 electrodes in a nitrogen atmosphere (7.28). Once again they found that some of the tubes were capped, and the shapes of these caps differed from those found on carbon nanotubes. For example, caps with flat tops were often seen, as illustrated in Fig. 7.11.

A quite different approach to the synthesis of boron nitride nanostructures has been described by Boulanger and colleagues from the Atomic Energy Research Centre at Saclay, France (7.29). These workers have developed methods for the synthesis of covalently bonded ceramic materials by laser-driven gas-phase reactions. To produce finely divided boron nitride by this technique, they reacted a mixture of ammonia and boron trichloride. The resulting material was examined both in its fresh state and following heat

Fig. 7.11. Boron nitride nanotube with flat end. Scale bar 5 nm. Courtesy Annick Loiseau.

treatment in nitrogen at temperatures up to 1650 °C. Concentric shelled particles similar to the carbon nanoparticles produced by arc-evaporation were observed, together with a variety of other structures including flat graphite-like platelets with edges folded over to avoid dangling bonds. Nanotube-like structures may also have been present, although images of these were not given.

Curled and closed boron nitride nanostructures have also been produced by electron irradiation of boron nitride inside an electron microscope. Taking a cue from Daniel Ugarte's work on the production of carbon onions (see next chapter), Florian Banhart and colleagues applied intense irradiation to small regions of crystalline BN (7.30). The effect of irradiation was to produce curling of the initially flat layers, resulting in some rounded, onion-like structures, as shown in Fig. 7.12. However, the curled BN structures observed by these workers were never found to be entirely closed. Subsequent studies by workers from Japan and France have shown that closed nanoparticles can be produced by irradiation of boron nitride in a TEM (7.31, 7.32). These BN 'fullerenes' could be either single-walled or multiwalled, and often had rectangular shapes reminiscent of the flat-topped BN nanotubes.

7.3.3 Structure of boron nitride tubes and nanoparticles

Boron nitride is isoelectronic with carbon, and hexagonal boron nitride (h-BN) has an sp^2-bonded structure very similar to that of graphite, as shown in Fig.

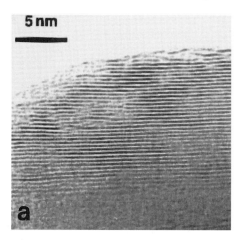 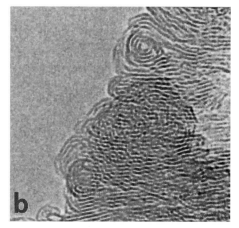

Fig. 7.12. Micrographs showing boron nitride before (a) and after (b) intense irradiation with 400 kV electrons (7.30).

7.13. Boron-containing fullerenes have been synthesised by Smalley's group (7.33) and, as noted in the previous section, fullerene-like BN nanoparticles have been produced using irradiation. However, it is unlikely that an exact C_{60} analogue containing only boron and nitrogen will be prepared because the formation of pentagonal rings would necessitate the creation of B–B or N–N bonds, which are much less energetically favourable than B–N bonds. On the other hand, four-membered B–N rings might be relatively stable, and may be present in BN nanotube caps, as discussed by Terrones and colleagues (7.27). These authors have proposed models of BN caps containing four-membered B–N rings, as shown in Fig. 7.14. These caps have the flat shapes which are often observed experimentally.

Cohen and colleagues have discussed the structural characteristics of BC_2N nanotubes (7.5). The number of possible structures for such tubes is greater than for all-carbon nanotubes, since there are several possible atomic configurations for a BC_2N sheet. Figure 7.15 shows the structure for two (4,4) armchair BC_2N tubes obtained by rolling up different atomic configurations. It will be noted that chirality is present in both structures, even though the corresponding all-carbon tubes would be achiral. It can also be seen that the two structures differ in that one has continuous chains of carbon atoms running helically around the surface while the other does not. In principle, BCN nanotubes should be more readily capped than pure BN tubes, since five-membered rings containing C, B and N should be relatively stable.

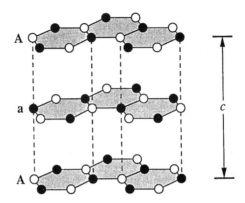

Fig. 7.13. The structure of hexagonal boron nitride (7.29). Closed circles, boron; open circles, nitrogen.

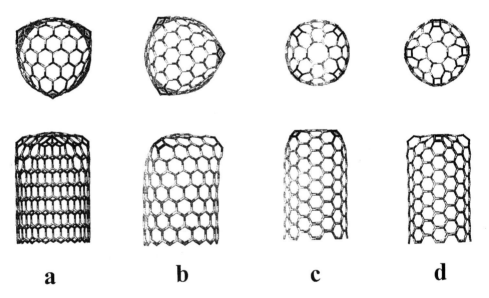

a   **b**   **c**   **d**

Fig. 7.14. Molecular models of BN nanotube caps (7.27). (a) and (b) show a cap containing three squares, while (c) and (d) show a cap containing four squares and one octagon.

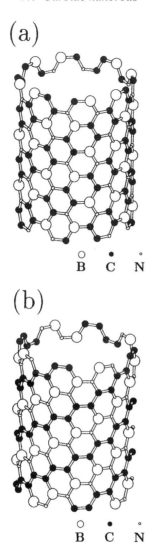

Fig. 7.15. Structures of two (4,4) armchair BC_2N nanotubes (7.5).

7.4 Carbide nanorods

The synthesis of nanoscale carbide rods by Lieber and colleagues involved reacting carbon nanotubes with volatile oxides and/or halides at temperatures around 1000–1500 °C (7.10, 7.34). The starting materials for these experiments were nanotubes produced catalytically rather than by arc-evaporation and were therefore free of nanoparticles and other unwanted material. In a typical experiment, a sample of tubes was reacted with the volatile oxide TiO at 1375 °C. The reaction products were solid rods rather than tubular structures, and tended to be straighter than the original nanotubes. Examples of TiC nanorods are shown in Fig. 7.16. Various morphologies were observed, including the regular sawtooth structures shown in Fig. 7.16(d), and the irregularly stepped structure shown in Fig. 7.16(e). The crystallographic directions were determined by electron diffraction and high resolution imaging. Nanorods were prepared of NbC, Fe_3C, SiC and BC_x using similar methods. Lieber and colleagues have measured the Young's modulus of SiC nanorods produced in this way, and found them to be comparable with the theoretical figure for single-crystal SiC.

Other groups have also explored the synthesis of nanorods using carbon nanotube precursors. It has been shown that nitride, as well as carbide nanorods can be formed in this way. For example, Chinese workers have synthesised gallium nitride nanorods by reacting Ga_2O vapour with NH_3 in the presence of carbon nanotubes (7.35). A group from Carnegie Mellon University have prepared tantalum carbide nanorods and nanoparticles and explored their superconducting properties (7.36). The potential of this method for producing nanoscale materials is substantial.

7.5 Discussion

The new inorganic nanostructures described in this chapter are among the more surprising byproducts of fullerene science. With the notable exception of Daedalus, few had foreseen the existence of stable, closed chalcogenide nanoparticles, and tubular variants of hexagonal boron nitride were certainly not envisaged before the discovery of the all-carbon analogues. Now that these materials have been synthesised, they seem to open up new avenues of research in structural inorganic chemistry. For example, the number of layered inorganic crystals is very large, so one might assume that the potential for synthesising further inorganic fullerenes is considerable. Indeed, Tenne's group, in collaboration with workers from Oxford, have recently prepared nickel dichloride cage structures and nanotubes (7.37). These may be the first of a family of nanostructures formed from layered halogen compounds. Certain oxides might also form fullerene-like particles. For example, triangular point defects in V_2O_5 could lead to closed structures; the work of Ajayan and co-workers has already shown that bent layers of V_2O_5 are stable (7.9). As far as BN-type inorganic nanotubes are concerned, these might also be synthesised from other III-V compounds, since several of these form layered crystals. The idea of using carbon nanotubes as templates for the synthesis of new inorganic structures also has considerable scope, and this line of research is still at a very preliminary stage.

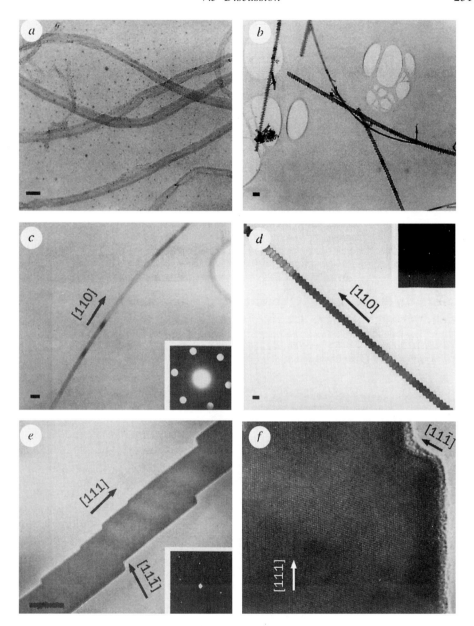

Fig. 7.16. Illustration of TiC nanorods synthesised by Lieber *et al.* (7.10). (a) Original carbon nanotubes, (b)–(e) typical nanorods, showing crystallographic directions, (f) high resolution micrograph showing atomic structure of nanorod.

Work on the properties of inorganic fullerenes and nanotubes is also still in its infancy, and in most cases applications seem a long way off. One possible exception to this is the idea of using WS_2 nanoparticles as solid lubricants, where they appear to show great promise. Little experimental work has been carried out on BN and BCN nanotubes, but Cohen *et al.* have carried out electronic structure calculations using the tight-binding method (7.4, 7.5). Pure boron nitride nanotubes were found to be semiconductors with band gaps which increased with tube diameter, rapidly approaching the h-BN value. It was also found that BN tubes could be either direct or indirect band gap semiconductors depending on helicity. BC_2N nanotubes should also be semiconducting, and Cohen *et al.* suggest that either p-type or n-type semiconductors could be obtained by changing slightly the stoichiometry of the tubes. The differences between the electronic properties of B and N-containing nanotubes and those of all-carbon tubes result largely from the ionic nature of the B–N bond. The stiffnesses of BN, BC and BCN nanotubes have also been discussed by theoreticians (7.38), who predict that these tubes should have moduli only slightly lower than those of pure carbon nanotubes (typically $\sim 900\,GPa$). Measurements of the vibration amplitude of BN nanotubes in a TEM appear to confirm that they have very high moduli (7.39).

References

(7.1) L. Pauling, 'The structure of the chlorites', *Proc. Nat. Acad. Sci. U.S.*, **16**, 578 (1930).
(7.2) R. Tenne, L. Margulis, M. Genut and G. Hodes, 'Polyhedral and cylindrical structures of tungsten disulphide', *Nature*, **360**, 444 (1992).
(7.3) R. B. Kaner, J. Kouvetakis, C. E. Warble, M. L. Sattler and N. Bartlett, 'Boron-carbon-nitrogen materials of graphite-like structure', *Mater. Res. Bull.*, **22**, 399 (1987).
(7.4) A. Rubio, J. L. Corkill and M. L. Cohen, 'Theory of graphitic boron nitride nanotubes', *Phys. Rev. B*, **49**, 5081 (1994).
(7.5) Y. Miyamoto, A. Rubio, M. L. Cohen and S. G. Louie, 'Chiral tubules of hexagonal BC_2N', *Phys. Rev. B*, **50**, 4976 (1994).
(7.6) O. Stephan, P. M. Ajayan, C. Colliex, Ph. Redlich, J. M. Lambert, P. Bernier and P. Lefin, 'Doping graphitic and carbon nanotube structures with boron and nitrogen', *Science*, **266**, 1683 (1994).
(7.7) N. G. Chopra, R. J. Luyken, K. Cherrey, V. H. Crespi, M. L. Cohen, S. G. Louie and A. Zettl, 'Boron nitride nanotubes', *Science*, **269**, 966 (1995).
(7.8) P. Gleize, M. C. Schouler, P. Gadelle and M. Caillet, 'Growth of tubular boron nitride filaments', *J. Mater. Sci.*, **29**, 1575 (1994).
(7.9) P. M. Ajayan, O. Stephan, Ph. Redlich and C. Colliex, 'Carbon nanotubes as templates for metal oxide nanocomposites and nanostructures', *Nature*, **375**, 564 (1995).
(7.10) H. J. Dai, E. W. Wong, Y. Z. Lu, S. Fan and C. M. Lieber, 'Synthesis and

characterisation of carbide nanorods', *Nature*, **375**, 769 (1995).

(7.11) E. J. W. Whittaker and J. Zussman, 'The characterisation of serpentine minerals by X-ray diffraction', *Mineral. Mag.*, **31**, 107 (1956).

(7.12) J. L. Hutchison, D. A. Jefferson and J. M. Thomas, 'The ultrastructure of minerals as revealed by high resolution electron microscopy', *Surf. Defect Prop. Solids*, **6**, 320 (1977).

(7.13) P. R. Buseck and J. M. Cowley, 'Modulated and intergrowth structures in minerals and electron microscope methods for their study', *Amer. Mineral.*, **68**, 18 (1983).

(7.14) D. R. Veblen and P. R. Buseck, 'Serpentine minerals: intergrowths and new combination structures', *Science*, **206**, 1398 (1979).

(7.15) K. Yada, 'Study of microstructure of chrysotile asbestos by high-resolution electron microscopy', *Acta Crystallogr. A*, **27**, 659 (1971).

(7.16) K. Wada, N. Yoshinaga, H. Yotsumoto, K. Ibe and S. Aida, 'High resolution electron micrographs of imogolite', *Clay Miner.*, **8**, 487 (1970).

(7.17) V. C. Farmer, M. J. Adams, A. R. Fraser and F. Palmieri, 'Synthetic imogolite – properties, synthesis, and possible applications', *Clay Miner.*, **18**, 459 (1983).

(7.18) L. Margulis, G. Salitra, R. Tenne and M. Tallanker, 'Nested fullerene-like structures', *Nature*, **365**, 113 (1993).

(7.19) M. Hershfinkel, L. A. Gheber, V. Volterra, J. L. Hutchison, L. Margulis and R. Tenne, 'Nested polyhedra of MX_2 (M = W, Mo; X = S, Se) probed by high resolution electron microscopy and scanning tunnelling microscopy', *J. Amer. Chem. Soc.*, **116**, 1914 (1994).

(7.20) Y. Feldman, E. Wasserman, D. J. Srolovitz and R. Tenne, 'High-rate gas-phase growth of MoS_2 nested inorganic fullerenes and nanotubes', *Science*, **267**, 222 (1995).

(7.21) R. Tenne, 'Fullerene-like structures and nanotubes from inorganic compounds', *Endeavour*, **20 (3)**, 97 (1996).

(7.22) L. Rapoport, Y. Bilik, Y. Feldman, M. Homyonfer, S. R. Cohen and R. Tenne, 'Hollow nanoparticles of WS_2 as potential solid-state lubricants', *Nature*, **387**, 791 (1997).

(7.23) Z. Weng-Sieh, K. Cherrey, N. G. Chopra, X. Blase, Y. Miyamoto, A. Rubio, M. L. Cohen, S. G. Louie, A. Zettl and R. Gronsky, 'Synthesis of $B_xC_yN_z$ nanotubules', *Phys. Rev. B*, **51**, 11,229 (1995).

(7.24) M. Terrones, A. M. Benito, C. Manteca-Diego, W. K. Hsu, O. I. Osman, J. P. Hare, D. G. Reid, H. Terrones, A. K. Cheetham, K. Prassides, H. W. Kroto and D. R. M. Walton, 'Pyrolytically grown $B_xC_yN_z$ nanomaterials: Nanofibres and nanotubes', *Chem. Phys. Lett.*, **257**, 576 (1996).

(7.25) K. Suenaga, C. Colliex, N. Demoncy, A. Loiseau, H. Pascard and F. Willaime, 'Synthesis of nanoparticles and nanotubes with well-separated layers of boron nitride and carbon', *Science*, **278**, 653 (1997).

(7.26) E. J. M. Hamilton, S. E. Dolan, C. M. Mann, H. O. Colijn, C. A. McDonald and S. G. Shore 'Preparation of amorphous boron-nitride and its conversion to a turbostratic, tubular form', *Science*, **260**, 659 (1993)

(7.27) M. Terrones, W. K. Hsu, H. Terrones, J. P. Zhang, S. Ramos, J. P. Hare, R. Castillo, K. Prassides, A. K. Cheetham, H. W. Kroto and D. R. M. Walton, 'Metal particle catalysed production of nanoscale BN structures', *Chem. Phys. Lett.*, **259**, 568 (1996).

(7.28) A. Loiseau, F. Willaime, N. Demoncy, N. Schramchenko, G. Hug, C. Colliex and H. Pascard, 'Boron nitride nanotubes', *Carbon*, **36**, 743 (1998).

(7.29) L. Boulanger, B. Andriot, M. Cauchetier and F. Willaime, 'Concentric shelled and plate-like graphitic boron nitride nanoparticles produced by CO_2 laser pyrolysis', *Chem. Phys. Lett.*, **234**, 227 (1995).

(7.30) F. Banhart, M. Zwanger and H.-J. Muhr, 'The formation of curled concentric-shell clusters in boron nitride under electron irradiation', *Chem. Phys. Lett.*, **231**, 98 (1994).

(7.31) D. Golberg, Y. Bando, O. Stephan and K. Kurashima, 'Octahedral boron nitride fullerenes formed by electron beam irradiation', *Appl. Phys. Lett.*, **73**, 2441 (1998).

(7.32) O. Stephan, Y. Bando, A. Loiseau, F. Willaime, N. Shramchenko, T. Tamiya and T. Sato, 'Formation of small single-layer and nested BN cages under electron irradiation of nanotubes and bulk material', *Appl. Phys. A*, **67**, 107 (1998).

(7.33) T. Guo, C. Jin and R. E. Smalley, 'Doping bucky – formation and properties of boron-doped buckminsterfullerene', *J. Phys. Chem.*, **95**, 4948 (1991).

(7.34) E. W. Wong, B. W. Maynor, L. D. Burns and C. M. Lieber, 'Growth of metal carbide nanotubes and nanorods', *Chem. Mater.*, **8**, 2041 (1996).

(7.35) W. Han, S. Fan, Q. Li and Y. Hu, 'Synthesis of gallium nitride nanorods through a carbon nanotube-confined reaction', *Science*, **277**, 1287 (1997).

(7.36) A. Fukunaga, S. Y. Chu and M. E. McHenry, 'Synthesis, structure, and superconducting properties of tantalum carbide nanorods and nanoparticles', *J. Mater. Res.*, **13**, 2465 (1998).

(7.37) Y. R. Hacohen, E. Grunbaum, R. Tenne, J. Sloan and J. L. Hutchison, 'Cage structures and nanotubes of $NiCl_2$', *Nature*, **395**, 336 (1998).

(7.38) E. Hernández, C. Goze, P. Bernier and A. Rubio, 'Elastic properties of C and $B_xC_yN_z$ composite nanotubes', *Phys. Rev. Lett.*, **80**, 4502 (1998).

(7.39) N. G. Chopra and A. Zettl, 'Measurement of the elastic modulus of a multi-wall boron nitride nanotube', *Solid State Commun.*, **105**, 297 (1998).

8

Carbon onions and spheroidal carbon

> Nevertheless, the most interesting question is whether, 500 years
> after Columbus reached the West Indies, flat carbon has gone the
> way of the flat Earth.
>
> Harry Kroto, *Nature*, 22 October 1992

When one browses through the carbon literature, spheroidal forms of carbon crop up again and again. Examples include carbon black, mesophase pitch and the graphitic particles in spherulitic graphite cast iron, all of which constitute technologically important materials. Despite many years of research, however, the structure and formation of these particles remains inadequately understood. Prior to the discovery of the fullerenes, models of spheroidal carbon particles had tended to involve assemblies of flat graphene fragments, which seems inherently unsatisfactory. The discovery of C_{60} provides us with a new paradigm for spheroidal carbon, and the time seems ripe to take a fresh look at spheroidal carbon in all its forms. This chapter will consider the evidence that fullerene-like elements may be present in the well-known forms of spheroidal carbon mentioned above. Firstly, though, a discussion is given of the most recently discovered form of spheroidal carbon: carbon onions.

8.1 Carbon onions

8.1.1 Discovery

The synthesis of carbon onions in 1992, like so many discoveries in fullerene science, was serendipitous. The Brazilian electron microscopist Daniel Ugarte, working at the Ecole Polytechnique Fédérale de Lausanne in Switzerland, had been examining carbon nanoparticles filled with gold and with lanthanum oxide, and decided to investigate the effect of electron irradiation on these structures (8.1). In the case of the encapsulated gold clusters, the irradiation caused the gold to be expelled from the carbon nanoparticle, while at the same time inducing a change in the nanoparticle's structure from faceted to rounded. Intrigued by the transformation of the carbon particles, Ugarte repeated the experiment with pure samples of cathodic soot. This time the effect was unmistakeable: under irradiation the nanotubes and nanoparticles evolved into almost perfect spheres apparently made up of concentric fullerenes, and Ugarte immediately despatched a paper to *Nature* (8.2). In this paper, Ugarte first used the term 'carbon onions' to describe the fullerene-like structures and, as can be seen in Fig. 8.1, one of his images made a memorable cover picture.

The appearance of Ugarte's paper created enormous interest and prompted much speculation about the potential bulk properties of this new form of carbon. Unfortunately, however, subsequent work has suggested that onions are rather unstable, and apparently only maintain their ideal structures when being imaged in an electron beam. In the absence of electron irradiation, onions seem to collapse into a much more disordered, though still spheroidal structure. This has not deterred a number of groups from pursuing research on these intriguing new structures in an attempt to understand their structure, stability and formation mechanisms.

8.1.2 Ugarte's experiments: irradiation of cathodic soot

In his *Nature* paper Ugarte described the irradiation of nanotubes and nanoparticles in a 300 kV transmission electron microscope. The irradiation was carried out by removing the condenser aperture and focusing the beam onto as small an area of sample as possible. Under these conditions he estimated the beam current to be of the order of $100–400\,A\,cm^{-2}$, which is 10–20 times the dose which would normally be used for imaging. After about 20 minutes of this treatment, the irradiated area had been almost exclusively transformed into onions. In fact, the high resolution images of onions which are included in Ugarte's paper appear somewhat disordered; only the beautiful image reproduced on the cover of *Nature* (Fig. 8.1) clearly shows the perfect concentric, quasi-spherical structure. There is some question about whether Ugarte was the first to observe carbon onions: images very similar to Ugarte's were published in 1980 by Sumio Iijima (8.3, 8.4), in a study of anomalous structures in evaporated carbon films. It appears that Iijima took this work no further, although Kroto and McKay carried out a detailed analysis of Iijima's images in 1988 (8.5).

Fig. 8.1. The cover of *Nature* dated 22 October 1992, showing Ugarte's micrograph of carbon onion.

8.1.3 *Production of onions from other carbons*

In Ugarte's experiments, the precursor for onion formation had been material extracted from the cathodic soot following arc-evaporation, i.e. mainly nanotubes and nanoparticles. It has since been demonstrated that several other forms of carbon can be transformed into onions in a similar way. Shortly after Ugarte's paper appeared, Edman Tsang and the present author, working in Malcolm Green's group in Oxford, showed that onions could be produced by irradiation of 'fullerene soot', the fullerene-containing carbon which con-

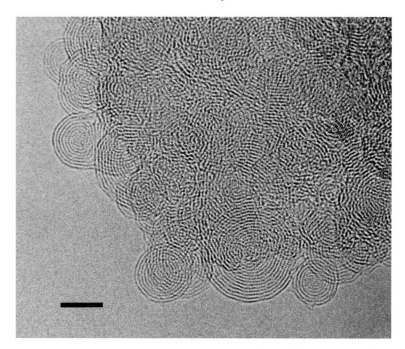

Fig. 8.2. Carbon onions prepared by intense electron irradiation of polymer-derived carbon (8.7). Scale bar 5 nm.

denses on the walls of the arc-evaporation vessel (8.6) . A group of onions produced in this way is shown in Fig. 8.2, while Fig. 8.3 shows two images of an individual onion taken a short time apart. Subsequent work has shown that onions can be made from a variety of conventional carbons (8.7, 8.8), as well as from diamond (8.9). It therefore seems that almost any form of carbon will evolve into onions when irradiated with a sufficiently intense electron beam inside an electron microscope.

It should be noted that individual fullerene molecules, as well as multi-layer carbon onions can be produced by electron irradiation. Thomas Füller and Florian Banhart have described *in situ* experiments in which various carbon materials were irradiated at high voltage while being simultaneously imaged using an on-line video system (8.10). Individual fullerene molecules could be seen to form at the surfaces of the carbons. Interestingly, the most commonly observed diameters for these molecules corresponded to those of C_{60} and C_{240}, which are expected to be among the most stable fullerenes. The smallest carbon cage observed in these studies apparently corresponded to either C_{28} or C_{32}. Adrian Burden and John Hutchison of the Materials Department at Oxford have also studied the formation of single-layer struc-tures, probably C_{60} molecules, in a controlled environment HREM (8.11).

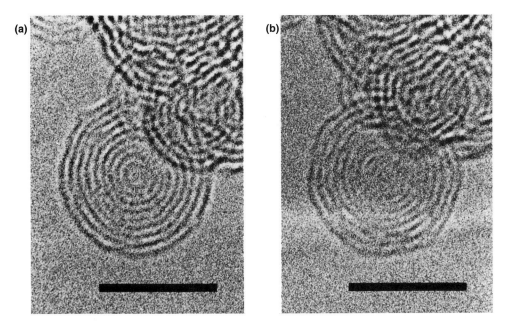

Fig. 8.3. Nine-shell carbon onion particle in irradiated fullerene soot sample. (a) Image showing circular profile, (b) image displaying slight faceting. Scale bar 5 nm.

Experiments of this kind may provide important new insights into fullerene formation.

As discussed in the previous chapter, the phenomenon of onion formation is not confined to carbon. Work by the Banhart group and others has demonstrated the formation of onion-like structures in irradiated boron nitride.

8.1.4 The structure of carbon onions

The *Nature* cover picture reproduced in Fig. 8.1, and the images shown in Figs. 8.2 and 8.3, illustrate the basic structural features of carbon onions. When fully formed, onions appear highly perfect in structure, with few obvious defects, and have a central shell about 0.7–1 nm in diameter, i.e. very close to the diameter of C_{60}. The onions are almost always quasi-spherical in shape, although they sometimes pass through a slightly faceted configuration, as shown in Fig. 8.3(b).

It is important to distinguish between carbon onions and carbon nanoparticles of the kind described in Chapter 3. Unlike onions, nanoparticles can have a wide variety of shapes, some polyhedral and some rounded, with central cavities of varying sizes, and they frequently contain defects. Moreover, all the electron microscopy evidence points to onions being comprised of concentric shells, whereas the structure of nanoparticles can be spiral, or nautilus-like, as well as concentric.

The detailed atomic structure of carbon onions has been the subject of much discussion. There are essentially two schools of thought. Some authors have assumed that the particles consist of nested fullerenes, as suggested by Kroto in the article which accompanied Ugarte's paper (8.12). A plausible model for onion structure is one made up of concentric 'magic number' (or Goldberg Type I) fullerenes. These fullerenes have N carbon atoms, where $N = 60b^2$, so that the first five are C_{60}, C_{240}, C_{540}, C_{960} and C_{1500} (with all fullerenes having the I_h symmetry). In an onion constructed from these fullerenes, the spacing between successive shells would be $\sim 0.34\,\text{nm}$, i.e. close to the interlayer spacing in graphite, and to the spacing observed experimentally for onions. A possible problem with this model is that most theoretical studies suggest that large fullerenes (C_{240} and larger) should be faceted, rather than spheroidal (8.5, 8.13). Although this faceting is only evident in certain directions, as shown in Fig. 8.4, one would expect some of the onions in experimental images to appear quite strongly faceted, rather than spheroidal. However, work by a Japanese group led by Kunio Takayanagi has suggested that the onions become preferentially aligned with the C_5 axis parallel to the beam, as a result of interaction with the magnetic field of the lenses, thus explaining their almost circular profiles (8.14). Detailed analysis of high resolution images by Zwanger and Banhart (8.15) has also provided support for the Goldberg-fullerene model.

Other workers have attempted to account for the sphericity of carbon onions by putting forward alternative structures (8.16–8.18). These have generally involved the introduction of heptagonal rings as well as pentagons into the structure of the carbon shells. The resulting structures are much more spherical than the corresponding fullerenes, as can be seen in Fig. 8.5. This shows a shell containing 1500 carbon atoms which contains 132 pentagons and 120 heptagons, taken from the work of Humberto and Mauricio Terrones (8.17). (It is fascinating to note the similarity of this structure to the radiolaria skeletons described by D'Arcy Thompson (8.19).) The sphericity of this structure suggests that it may represent a more realistic model for the outer shells of carbon onions than the 'perfect fullerene' model.

8.1.5 Formation mechanism of carbon onions

In a paper in *Chemical Physics Letters* in 1993 (8.20) Ugarte published a beautiful series of high resolution micrographs showing the transformation of a nanoparticle into an onion; these are reproduced in Fig. 8.6. These images

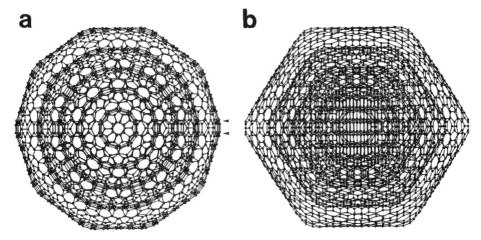

Fig. 8.4. Two views of a five-shell carbon onion with the 'concentric fullerene' structure, along (a) C_5 and (b) C_2 symmetry axes (8.14).

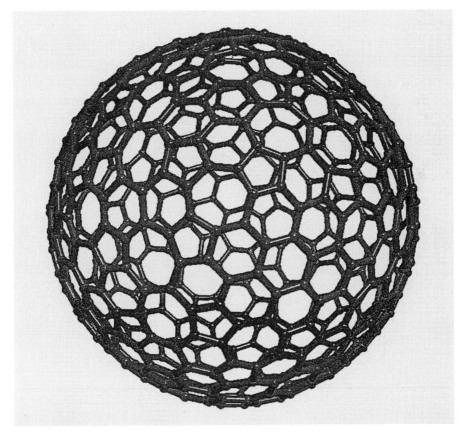

Fig. 8.5. The Terrones model of carbon onion structure, incorporating pentagon–heptagon patches (8.17).

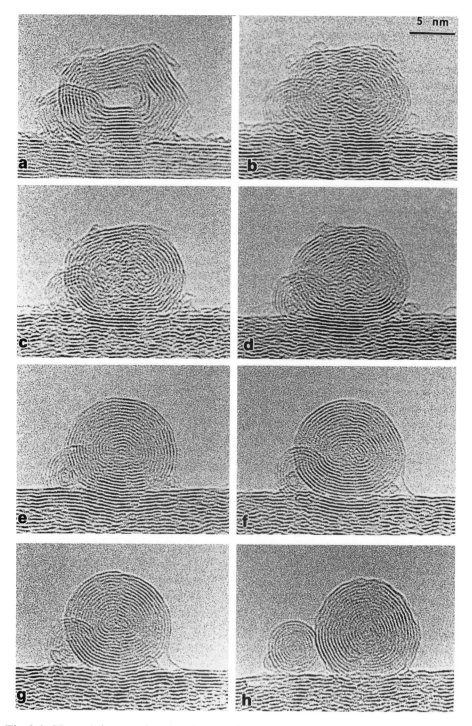

Fig. 8.6. Ugarte's images showing the transformation of a carbon nanoparticle into an onion (8.20).

show that the initial stage of the transformation involves a partial collapse of the nanoparticle's structure, resulting in the elimination of the hollow cavity at the centre (Fig. 8.6(b)). Next, regraphitization of some of the surface layers occurs (Fig. 8.6(c) and (d)), and the shape of the particle becomes more ellipsoidal. Gradually the shape evolves into a sphere (Fig. 8.6(e)–(h)), with the concentric shell structure becoming more and more perfect. This latter process seems to occur from the outside in, and has been described by Ugarte as 'internal epitaxial growth'. A schematic representation of the process, also taken from Ugarte's work (8.21), is shown in Fig. 8.7.

A detailed series of studies of the formation mechanism of onions has been carried out by Florian Banhart and colleagues (8.8, 8.22). These workers irradiated a number of different carbon materials at electron energies in the range 100–1250 kV, and demonstrated that the time required to transform the materials completely into onions decreased with increasing energy (8.8). From this they concluded that electronic excitations are unlikely to be responsible for the transformation into onions, (at least at energies above 200 kV) since these become progressively less likely with increasing electron energy. They also ruled out the idea that a simple heating effect is responsible for the transformation, since the temperatures experienced by the carbon materials are unlikely to exceed 200–300 °C (8.22). It should also be noted that high-temperature heat treatment of carbon materials such as fullerene soot does not transform them to onions, but rather to nanoparticles and tubes (see p. 25). Banhart and colleagues have put forward a mechanism for the transformation of nanoparticles into onions which involves the sputtering of atoms out of their original positions by knock-on collisions with electrons from the beam, followed by a shrinkage of the shells. This results in a contraction of the whole particle, which tends to force it into a spherical configuration. This mechanism differs markedly from the one proposed by Ugarte.

The transformation of carbon nanoparticles into onions has also been discussed by Humberto and Mauricio Terrones (8.17). They suggest that the initial phase involves the destruction of strained regions, which include five-membered rings, to produce 'fullerenes with holes' which are relatively flexible, and tend towards spheroidal shapes. The holes are expected to be rapidly filled with carbon atoms in five- and seven-membered rings to produce structures of the type discussed above (Fig. 8.5).

Clearly there is no general agreement on the formation mechanism of carbon onions, and further experimental work is needed to resolve this issue.

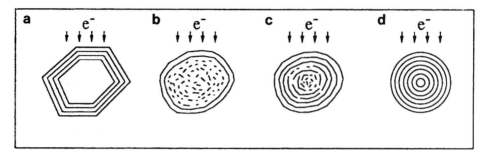

Fig. 8.7. Schematic illustration of Ugarte's proposed mechanism of onion formation (8.21).

8.1.6 Stability of carbon onions

A group from Bologna were among the first to demonstrate that carbon onions are unstable when not being irradiated in an electron beam (8.23). In a paper published in 1995 they showed that a current density of approximately $150\,A\,cm^{-2}$ was required to maintain the perfect onion structure and avoid structural collapse. Further detailed work on the decay of carbon onions was carried out by Banhart and colleagues (8.8). These workers found that the rate of decay could vary considerably, ranging from 10 minutes after the removal of irradiation, to several months. The speed of decay appeared to depend on the nature of the surrounding material, and appeared to occur from the outside to the inside. A series of micrographs illustrating this decay is shown in Fig. 8.8.

Theoretical work on the stability of carbon onions is difficult at present because of the uncertainty concerning their precise structure, and detailed calculations on the stability of large multilayer carbon onions are not yet available.

8.1.7 Bulk synthesis of carbon onions

The apparent instability of carbon onions under normal conditions suggests that bulk synthesis of carbon particles with the perfect onion structure may never be possible. Nevertheless, a number of groups have explored methods of producing onions, or onion-like structures, in bulk, and these will now be summarised.

In 1994 a Russian group published two papers describing the apparent production of onion-like carbon from ultra-disperse diamond (8.24, 8.25). Vladimir Kuznetsov and colleagues had previously found ultra-disperse diamond (average particle diameter $\sim 4.5\,nm$) in detonation soot. When they

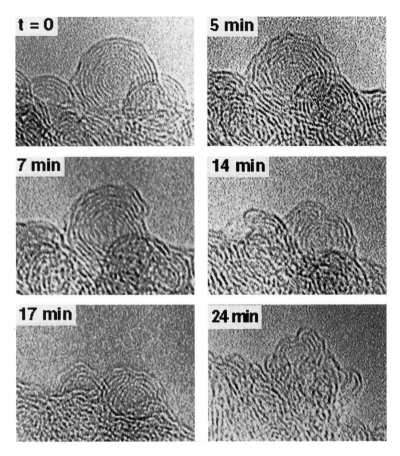

Fig. 8.8. Micrographs showing decay of carbon onions, courtesy of Florian Banhart.

heat-treated this material at 1000–1500 °C they found that it was transformed into a graphitic material made up almost entirely of 'onion-like' particles. However, the onions were generally less spherical and much less perfect than Ugarte's, and it is not clear whether the particles were stable in the absence of an electron beam.

In the following year a French group described a quite different method for generating spheroidal carbon structures which they described as 'fullerene onions' (8.26). Their technique involved directing beams of carbon ions onto copper and silver substrates at high temperature. The effect of this treatment is to implant carbon ions into the copper, which then diffuse towards the surface and form a variety of carbon structures, depending on the conditions used. The onions formed in this way could be as large as several micrometres in diameter. High resolution electron micrographs of the particles indicated that they had structures similar to those of Ugarte's onions. However, it is not clear whether they retain this highly perfect structure outside the electron microscope. Subsequently, refinements to their synthesis method have been described (8.27, 8.28).

Other groups have described new methods of producing small spheroidal carbon particles. For example, workers from the Georgia Institute of Technology used a catalytic process to produce carbon spheres from natural gas (8.29). These new processes may be technologically interesting as routes to pure spheroidal carbon particles with relatively narrow size distributions.

8.1.8 The formation of diamond inside carbon onions

In the previous section the conversion of highly dispersed diamond into onion-like particles by Kuznetzov and colleagues was described. Florian Banhart and Pulickel Ajayan reported in 1996 that the reverse process can also occur (8.30–8.32). These workers were investigating the behaviour of carbon onions when subjected to simultaneous annealing and irradiation inside an electron microscope. They found that heating the onions to 700 °C while irradiating them with an intense electron beam resulted in the formation of small domains of diamond inside the onions. This is illustrated in the beautiful image shown in Fig. 8.9.

In their initial work (8.30), Banhart and Ajayan carried out the irradiation inside a TEM, thus producing only tiny amounts of diamond. It was subsequently shown that much higher yields of diamond could be formed by irradiating with ions from an accelerator rather than with electrons (8.32). These studies might lead to new routes for the synthesis of nanoscopic diamond particles, and, at the same time, to a better understanding of the graphite–diamond transition.

8.2 Spheroidal carbon particles in soot

8.2.1 Background

As noted above, quasi-spherical carbon particles are an important component of soot. In most soots, these particles are accompanied by a poorly defined mixture of ash, hydrocarbons and inorganic material. However, when formed under carefully controlled conditions, samples of soot consisting almost entirely of carbon particles can be produced, and this material, known as carbon black, is manufactured industrially on a large scale, primarily for use as a filler in rubber products. It is also used as a pigment, an application which dates back to prehistoric times, and as a component of xerographic

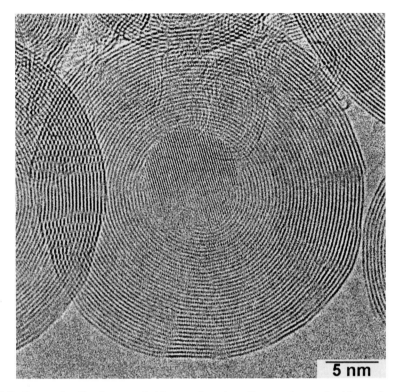

5 nm

Fig. 8.9. Image by Banhart and Ajayan showing diamond core inside carbon onion particle (8.30).

toners. A variety of industrial processes are used, the most important of which is the 'furnace black process', which involves the partial combustion of petrochemical or coal tar oils. Other methods for producing carbon black involve gas phase pyrolysis rather than combustion. Depending on the method of production, carbon blacks with particle diameters ranging from about 10 nm to about 500 nm can be made. A transmission electron micrograph of typical carbon soot particles at intermediate magnification is shown in Fig. 8.10, while a higher magnification image is shown in Fig. 8.11. This high resolution image shows the individual curved graphene layers which make up the spheroidal particles. In the following two sections an attempt will be made to summarise 'pre-fullerene' thinking on the growth and structure of spheroidal carbon soot particles. The icospiral mechanism of soot particle growth, put forward by Kroto, Smalley and their collaborators will then be discussed, and evidence for fullerene-like elements in soot and carbon black particles evaluated.

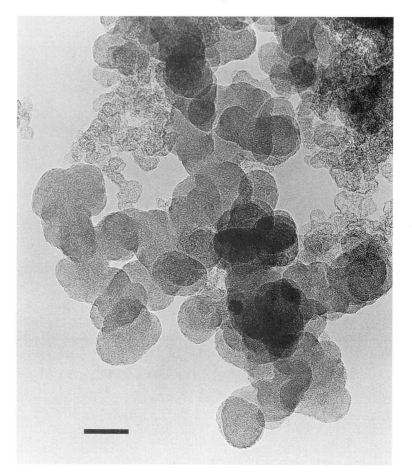

Fig. 8.10. Micrograph of commercial carbon black particles (Vulcan XC-72: Cabot Corporation). Scale bar 50 nm.

8.2.2 Growth mechanisms: the traditional view

Research into soot formation can be traced back to the work of Humphrey Davy and Michael Faraday at the Royal Institution in the nineteenth century. It was Davy who originally ascribed the yellow incandescence of a flame to glowing carbon particles, while Faraday, in his famous series of lectures on 'The chemical history of a candle', used a burning candle as the starting point for a wide-ranging dissertation on natural philosophy (8.33). Despite the vast amount of work which has been carried out on the subject since that time, the mechanism of carbon black formation is still a matter for debate. A detailed discussion of the various theories would be beyond the scope of the present work; a number of comprehensive reviews have been published (8.34–8.36).

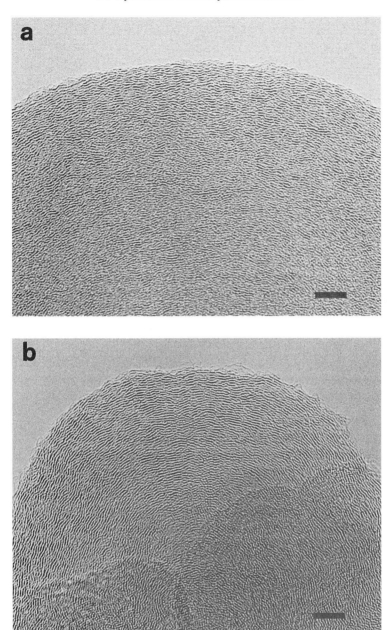

Fig. 8.11. High resolution electron micrographs showing individual graphene layers in commercial carbon black particles, courtesy Adrian Burden. (a) Raven 430 (Columbian chemicals company), (b) Vulcan XC-72. Scale bar 5 nm.

However, a brief summary can be given of the basic aspects of soot formation as they are currently understood.

It is believed that soot particles form in three distinct phases. The first stage, known as *particle inception*, involves homogeneous reactions between hydrocarbon species which combine into larger aromatic layers and eventually condense out of the vapour phase to form nuclei. In the second stage, called *growth*, two processes occur: nuclei coalescence and surface deposition, the latter process being responsible for most of the mass increase of the primary particles. Finally, in the *chain formation* stage, relatively large spheroidal particles become joined together, without coalescing, to form long chains.

Many aspects of this mechanism are poorly understood, in particular the nature of the initial nucleus. The most widely accepted view is that the nuclei are liquid microdroplets, thus explaining the particles' sphericity. However, there is much uncertainty about the mechanism whereby the droplets transform into solid graphitic particles.

8.2.3 *The icospiral growth mechanism*

Harry Kroto, Rick Smalley and their co-workers became interested in the subject of soot formation in the aftermath of their discovery of buckminsterfullerene in 1985. They were well aware of the uncertainty in the carbon community concerning the initial stages of soot formation, and naturally enough wondered whether fullerene-like structures might be involved. After all, the carbon particles in soot, like fullerenes, were spheroidal in form, and both result from the condensation of carbon species from the gas phase. In late 1985 they submitted a paper to the *Journal of Physical Chemistry* in which they proposed a mechanism for soot nucleation based on the 'pentagon-road' model of fullerene formation (8.37). This mechanism was later refined by Kroto and McKay (8.5). The essential element of the pentagon-road model is the incorporation of pentagonal rings into a growing carbon network, driven by the need to eliminate dangling bonds (see also p. 16). If the pentagons occur in the 'correct' positions then C_{60} and other fullerenes will result, but in general closed structures will not be formed. Kroto and Smalley suggested that if the growing shell fails to close, it would tend to curl around on itself like a nautilus shell, as illustrated in Fig. 8.12. In the refined model (8.5), Kroto and McKay suggested that the pentagons in a new layer would tend to form directly above those in the previous layer in an epitaxial manner, resulting in 12 columns of pentagonal rings. They argued that the spiralling structure would become increasingly faceted as it grew larger, in the same way that giant fullerenes are expected to be much more faceted than small ones. In support of these ideas, Kroto and

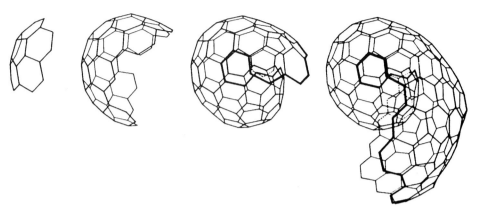

Fig. 8.12. Illustration of icospiral nucleation model (8.5).

McKay carried out a detailed analysis of Sumio Iijima's 1980 images of spheroidal carbon particles in evaporated carbon films. They showed that these images were consistent with their icospiral model and displayed clear evidence of faceting, indicating the presence of columns of pentagonal rings.

It would probably be an understatement to say that the icospiral nucleation model was not received enthusiastically by the soot community. Two soot specialists in particular, Michael Frenklach of Pennsylvania State University and Lawrence Ebert of Exxon, argued strongly against the new theory (8.38, 8.39). According to some accounts, their hostility was partly based on scepticism about the very existence of C_{60} (8.40, 8.41), although this has been rejected by Ebert (8.42), who also strenuously denies ever having likened fullerene science to cold fusion! In fact, the arguments of Frenklach and Ebert were based on a thorough analysis of both the kinetics of soot formation and the structural characteristics of the particles. Concerning kinetics, Frenklach and Ebert used computer simulations to show that the growth of shell structures would be much slower than those of planar fragments. Soot formation is known to be extremely rapid, so it seemed unlikely that shell growth could be involved. Their structural arguments were based on X-ray diffraction and ^{13}C NMR patterns of combustion soot, which they suggested were inconsistent with the icospiral model. They pointed out that d-spacings for continuously curving icospirals would be lower than those observed experimentally in XRD studies of soot, while ^{13}C NMR spectra of soot resemble those of aromatic molecules much more closely than those of fullerenes. However, evidence from XRD and ^{13}C NMR of soot particles is difficult to interpret when one is dealing with disordered materials such as carbon blacks, and cannot be said to provide definitive proof of the structure. Moreover, Frenklach and Ebert do not propose a detailed alternative to the icospiral nucleation model which would explain the spheroidal shape of combustion soot particles. It would therefore be wrong to dismiss the idea that a fullerene-like growth mechanism, involving the incorporation of pentagons into growing sheets, might be involved in soot formation. However, the picture of a *continuously* growing sheet of the type envisaged by Kroto and Smalley may have to be modified. As discussed below, recent studies of carbon black structure using scanning tunnelling microscopy suggest that the particles consist of overlapping curved scales, rather than a seamless nautilus-like shell.

8.2.4 The structure of carbon black

A huge volume of research has been carried out on carbon blacks (8.36), and only a brief summary can be given here. The first X-ray diffraction studies were carried out by Warren in 1934 (8.43), who demonstrated the presence of layer planes of graphite-like carbon. However, no evidence of the three-dimensional graphite structure was detected. Many subsequent X-ray diffraction studies have been carried out (e.g. 8.44), enabling the dimensions of the individual crystallites to be determined. The two parameters used to characterise the structure of disordered carbons are L_a, the length of the fragment in the in-plane direction and L_c the length in the c direction. For carbon blacks, L_a usually falls in the range 1–3 nm, while L_c is usually of the order of 1.5 nm. However, these values probably underestimate the size of the sheets which make up carbon black, as discussed below.

While X-ray diffraction could reveal the nature of the basic structural units, detailed information on the internal organisation of the carbon black particles had to await the development of high resolution electron microscopy. In the late 1960s, work by Heidenreich, Hess and Ban (8.45) revealed that the graphene crystallites were arranged concentrically, and this led them to propose the model shown in Fig. 8.13. This model has been very widely reproduced, and is still apparently regarded by many workers in the field as essentially correct. However, high resolution electron microscope images can be extremely difficult to interpret (8.46), and it is unwise to put too much faith in a model obtained by HREM alone. For example, the contrast displayed by a graphene sheet depends critically on its orientation with respect to the beam, so that the individual sheets making up the carbon black particle may be much longer than they appear to be. Recent work using scanning tunnelling microscopy has suggested that the curved sheets may indeed be much more extensive than previously envisaged. Jean-Baptiste Donnet and Emmanuel Custodéro have presented STM measurements apparently indicating that carbon black particles are made up of overlapping scales (8.47, 8.48), as illustrated in Fig. 8.14.

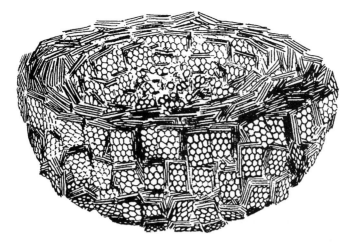

Fig. 8.13. Hess–Ban–Heidenreich model of carbon black structure (8.45).

Fig. 8.14. Donnet–Custodéro model of carbon black structure (8.47, 8.48).

This model is reminiscent of the Kroto–McKay icospiral growth model, and Donnet and Custodéro believe that their observations provide support for the view that the 'seed' for soot particle growth may be a fullerene-like icospiral.

There are other reasons for believing that carbon black and soot particles may have a fullerene-like microstructure. For example, the observation that fullerenes can form during the combustion of hydrocarbons proves that pentagon-containing structures can form in flames. Fullerenes in flames were first observed by Klaus Homann and co-workers at the Max Planck Institute for Physical Chemistry in Darmstadt in 1987. This group reported finding traces of C_{60} in 'sooting' acetylene–oxygen and benzene–oxygen flames, i.e. flames in which there is insufficient oxygen for complete combustion (8.49). Four years later, a group led by Jack Howard at the Massachusetts Institute of Technology showed that *macroscopic* amounts of fullerenes could be produced in sooting benzene–oxygen flames (8.50).

Further evidence in support of the view that carbon black and soot particles may be fullerene-like comes from a consideration of the structures formed when these materials are heated at very high temperatures, i.e. 2500 °C and above (8.51). It is well established that such treatments transform carbon black particles into faceted particles which sometimes appear to have closed shell-like structures. The precise structure of the graphitized particles depends on the nature of the original carbon black. In some cases, relatively large, discrete particles are formed, as shown in Fig. 8.15(a), taken from the work of Graham and Kay (8.52). Other graphitized carbon blacks have a less well-defined structure as in Fig. 8.15(b), from the work of Marsh and colleagues (8.53), with many bent and faceted layer planes and some apparently closed shell structures. The presence of sharply bent planes and closed particles is indicative of the presence of pentagonal rings, and suggests that fullerene-like elements may have been present in the original carbon black and soot particles. Further detailed work on the graphitization of carbon blacks might help to confirm this.

8.3 Spherulitic graphite cast iron

8.3.1 History

The history of iron–carbon alloys extends back at least 3000 years. They were probably first produced through the accident of iron-bearing stones being exposed to the hot charcoal in a fire. The action of the charcoal would have converted the stones to spongy pieces of iron which could then be recovered from the fire. Eventually the technology was developed for the forging, welding and working of iron into useful objects. Until the eighteenth century, virtually all iron-smelting furnaces were fuelled by relatively scarce charcoal. Around 1709, the English iron worker Abraham Darby found a way of making cast iron using plentiful coke (purified coal). This innovation was one of the factors leading to the birth of the Industrial Revolution.

In cast iron, the carbon content ranges from around 2% to 5% by weight. Other elements are also present, principally silicon. The nature of the carbon in

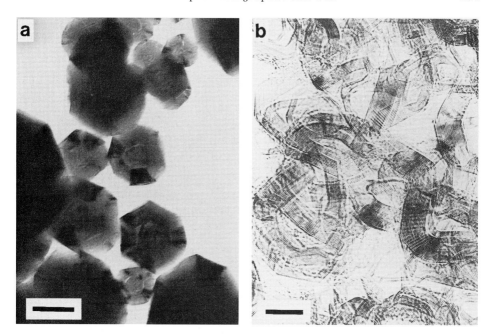

Fig. 8.15. (a) Faceted particles in graphitized carbon black sample (8.52). Scale bar 100 nm. (b) Graphitized carbon black containing bent and faceted layer planes, and some closed particles (8.53). Scale bar 10 nm.

the alloy depends on how the carbon comes out of solution, which in turn depends on the cooling rate and silicon content. Slow cooling and high silicon leads to the formation of 'grey iron' in which the precipitated carbon is in the form of flakes of graphite typically 10–100 μm in length. These flakes have very little strength and with their sharp ends they act like numerous cracks within the microstructure. One result of this is that grey iron has very little tensile strength, but it can be utilised usefully in compression. Indeed during the early part of the nineteenth century cast iron beams were used widely for bearing compressive loads in buildings.

If carbon could be persuaded to precipitate as compact nodules in place of flakes then this would enormously increase the range of application of cast iron. A method for achieving this, known as 'malleablising' was introduced in the nineteenth century. For this process the composition and casting conditions are chosen such that during the solidification of cast iron the carbon comes out of solution, not as graphite, but as cementite, the hard white iron carbide Fe_3C. The 'white iron' casting is then malleablised, i.e. given a prolonged anneal at sub-eutectic temperatures which results in the cementite decomposing and, depending on the annealing temperature, quite compact graphite nodules are formed. This malleablised cast iron is much tougher than

conventional grey iron and the product found extensive application throughout the nineteenth century and the first half of the twentieth century.

Malleablising is, however, an expensive process and clearly it was desirable
to develop a composition or procedure which would produce a cast iron in
which, in the as-cast condition, the graphite consisted of compact near-spheri-
cal nodules. This breakthrough was achieved in the late 1940s when it was
discovered by Williams and Morrogh (8.54) and others that the addition of
rare earth metals or magnesium immediately before pouring the casting did
indeed produce the desired result – the graphite phase appearing in the
microstructure not as flakes but as compact spherulites. This was one of the
most significant discoveries in the field of casting in the twentieth century.
Known as spheroidal graphite (SG) cast iron, this alloy can be used in
engineering, particularly in the automobile industry, and in construction.

8.3.2 *The structure of spherulitic graphite*

An optical micrograph of some typical graphite spherulites is shown in Fig.
8.16. The particles here range from about 5 μm to 25 μm in diameter; in some
cases particles can grow as large as 100 μm. They are thus far larger than typical
soot particles. A higher magnification optical micrograph of an individual
spherulite, recorded with polarised light, is shown in Fig. 8.17. This shows that
the structure of the spherulite is not onion-like but instead consists of closely
packed and intertwined filamentary segments growing out from a central
nucleus. Electron microscopy and diffraction show that the radial segments
grow primarily in the *c*-direction, so that the basal planes remain approximately parallel to the surface of the growing particle. However, the detailed structure
of the segments is not known; this will be discussed further below.

The nature of the initial nucleus for SG growth is also unknown. The
possibility that fullerene-like structures might be involved in the nucleation of
SG has been discussed by a number of groups (8.56–8.60). The most detailed
discussion has been given by Double and Hellawell (8.56, 8.57) who propose a
model for SG growth similar to the McKay–Kroto icospiral nucleation model
(although they are vague about the precise details). They point out that
spheroidal growth is the only mechanism which allows large graphitic particles
to form by continuous addition to a growing sheet. The formation of flake-like
graphite, made up of stacks of layer planes, would require repeated nucleation
events (unless the graphene sheets folded back on themselves), and therefore a
greater activation energy.

It has been suggested that the involvement of fullerenes at the nucleation
stage might explain the role of foreign elements in promoting SG growth. The

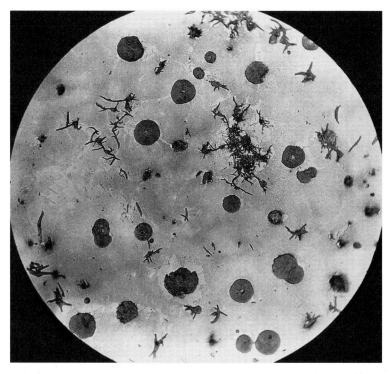

Fig. 8.16. Optical micrograph showing both spherulitic and flake-like graphite structures in magnesium-treated cast iron (8.55).

traditional view is that these elements increase the interfacial energy between the carbon and the matrix, favouring spheroidal precipitates, but this is by no means certain. Lanthanum, which is one of the rare earth metals used in promoting SG growth, is readily incorporated into fullerene cages (see p. 156). This prompted the idea that a metallofullerene might be the seed for spherulite growth (8.58–8.61). However, it should be borne in mind that spherulitic graphite does not necessarily require the presence of additives, but can also form in a 'clean' melt, from which most impurities have been removed.

To return to the structure of the filamentary segments, this has been discussed by a number of groups (e.g. 8.56, 8.57). Double and Hellawell (hereafter DH) suggest that the filamentary segments consist of helical cones, as illustrated in Fig. 8.18. In support of this idea, DH note that graphitic whiskers with a rather similar structure can be produced by pyrolysis of carbon monoxide on a silicon carbide substrate (8.62). They point out that the lowest energy configuration of such a cone-helix will occur when there is a favourable coincidence configuration between the growing sheet and the underlying one, which in turn implies that there will be certain favoured cone apex angles. One

Fig. 8.17. Optical micrograph showing filamentary structure of individual graphite spherulite in SG iron (8.55).

of these favoured angles will be 140°, which happens to be precisely the angle observed in the graphitic whiskers referred to above. This case is illustrated in Fig. 8.19. Double and Hellawell describe the defect at the apex as a screw dislocation, but do not describe the atomic structure of this dislocation in detail. One possibility which DH do not seem to have considered is that there might be an individual pentagon at the apex. This would result in a cone angle of 112.9°. Of course, in the absence of any other defects, a single pentagon would simply result in a seamless cone, but the introduction of an imperfection such as a foreign atom from the matrix might produce a discontinuity in the growing sheet, resulting in overgrowth, and a spiralling filament of the kind envisaged by DH.

Very little experimental work has been carried out on the possible role of fullerenes in SG growth. Double and Hellawell have looked for evidence of fullerenes in the graphitic residues left after chemical and electrolytic dissolution of SG iron, but with no success.

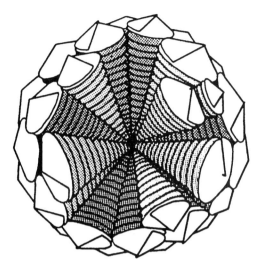

Fig. 8.18. Sketch of cone-helix model of the structure of spherulitic graphite (8.56).

8.3.3 The precipitation process

Whether or not the graphite crystallises in a spherulitic form depends on a variety of factors. Of prime importance is the precise nature of the precipitation process. In un-inoculated hypoeutectic cast iron (i.e. the common form of grey iron) the first phase to precipitate on cooling below the liquidus is austenite, with carbon coming out of solution and forming graphite flakes at the eutectic temperature. These flakes have the appearance of being an interlocking semi-continuous phase, so during their growth they will always be in contact with the liquid phase and can be fed with carbon directly from it (so the flakes will be largely free of stress).

In contrast, at the very early stages of formation of a spherulite in an inoculated hypoeutectic alloy, the nucleus and its accretions will be surrounded by a shell of solid austenite and continued growth can only occur by interstitial diffusion of carbon through the austenite. Of course, during growth the graphite will be occupying space hitherto occupied by iron atoms and the provision of this space may be difficult, giving rise to a large hydrostatic compressive force. Such a force is equivalent to having a high interfacial energy and will promote spherical growth. It need hardly be stressed that this is a complex area, and further discussion would be beyond the scope of the present book.

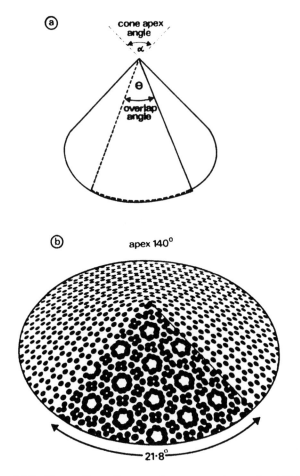

Fig. 8.19. Formation of a cone from a graphene sheet (8.56).

8.4 Spheroidal structures in mesophase pitch

A discussion of spheroidal forms of carbon would not be complete without a mention of mesophase pitches. It should be stressed at the outset, however, that there is currently no evidence that the spherical structures observed in these materials are fullerene-like.

Pitches are solid carbonaceous materials derived from precursors such as coal tar and petroleum by heat treatment or distillation, and contain molecules with a wide range of structures and sizes (8.63, 8.64). They are of great commercial importance since, when heated to temperatures of the order of 2500–3000 °C under an inert atmosphere, they can be converted into single-crystal graphite (unlike most carbons derived from solid materials, which tend to be non-graphitizing (8.51)). The resulting synthetic graphites are employed

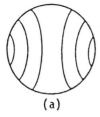 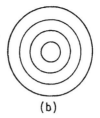

(a)   (b)

Fig. 8.20. Schematic illustration of structure of mesophase pitch particles (8.63). (a) The usual 'Brooks and Taylor' (8.64) structure, with equatorial stacking of lamellae. (b) Less common onion-like structure.

in such diverse applications as electrodes, nuclear moderators and carbon fibres.

The transformation of a pitch into crystalline graphite involves a number of distinct stages. On heating, most pitches melt to give isotropic fluids. At temperatures above about 375 °C, alignment of lamellar molecules occurs leading to the formation of a mesophase or liquid crystal system containing spheroidal structures. As the growth continues, the isotropic phase is exhausted and the spherulites coalesce. These spherulites usually have a structure in which small lamellar fragments are equatorially stacked, as illustrated in Fig. 8.20(a), although occasionally concentric structures are observed as in Fig. 8.20(b). Further heat treatment of the pitch results in the formation of a coke, that is a high carbon-content solid made up of small graphite-like layers but with no long-range order. Crystalline graphite is then produced by treatment of the coke at temperatures in the range 2500–3000 °C.

Kroto and colleagues have considered the possibility that mesophase pitch particles may be fullerene-like in structure (8.60, 8.61), noting the similarity of the concentric structure to that of carbon onions. However, the fact that cokes formed from pitch, unlike carbon blacks, form single-crystal graphite on high-temperature heat treatment seems to argue against a fullerene-like structure.

8.5 Discussion

Ugarte's beautiful 1992 paper describing the synthesis of carbon onions generated the same kind of excitement that Iijima's paper on carbon nanotubes had produced a year earlier. However, much of this enthusiasm evaporated when it was found that the onions appeared to be unstable outside the highly artificial environment of an electron microscope. Nevertheless, fundamental research on onions will undoubtedly continue, since many questions remain concerning these mysterious objects, not least about their precise structure. The remarkable work of Banhart and Ajayan on nanoscopic diamond formation inside carbon onions provides a further reason for pursuing onion research.

The instability of carbon onions might suggest that Harry Kroto was premature in writing the obituary of flat graphite. However, the important point is that the discovery of structures like fullerenes and carbon onions forces us to think again about the nature of well-known forms of carbon. Flat graphite undoubtedly exists and is, as far as we know, the most stable form of bulk carbon at ordinary temperatures and pressures. But the fact remains that many carbons cannot be transformed into crystalline graphite, even at temperatures of 3000 °C and above. Recent work suggests that this is because the so-called non-graphitizing carbons form closed, fullerene-like particles on heat treatment, rather than flat graphite (8.51). However, these particles resemble small carbon nanoparticles rather than carbon onions.

As far as spheroidal soot and carbon black particles are concerned, the idea that fullerene-like elements may be present in these particles is still controversial. It has been argued in this chapter that the transformation of carbon black particles into apparently fullerene-related nanoparticles following high-temperature heat treatments provides evidence that fullerene-like elements were present in the original particles, but this certainly cannot be considered definitive proof. A great deal of further structural characterisation of soot and carbon black particles will be required before this important issue can be resolved.

More work is also needed on the other materials discussed in this chapter: mesophase pitch and spherulitic graphite in cast iron. Although these materials may not be fullerene-related, their detailed structures remain unknown.

References

(8.1) D. Ugarte, 'How to fill or empty a graphitic onion', *Chem. Phys. Lett.*, **209**, 99 (1993).

(8.2) D. Ugarte, 'Curling and closure of graphitic networks under electron-beam irradiation', *Nature*, **359**, 707 (1992).

(8.3) S. Iijima, 'Direct observation of the tetrahedral bonding in graphitized carbon black by high-resolution electron microscopy', *J. Cryst. Growth*, **50**, 675 (1980).

(8.4) S. Iijima, 'High resolution electron microscopy of some carbonaceous materials', *J. Microscopy*, **119**, 99 (1980).

(8.5) H. W. Kroto and K. McKay, 'The formation of quasi-icosahedral spiral shell carbon particles', *Nature*, **331**, 328, (1988).

(8.6) S. C. Tsang, P. J. F. Harris, J. B. Claridge and M. L. H. Green, 'A microporous carbon produced by arc-evaporation', *J. Chem. Soc., Chem. Commun.*, 1519 (1993).

(8.7) P. J. F. Harris, S. C. Tsang, J. B. Claridge and M. L. H. Green, 'High-resolution electron microscopy studies of a microporous carbon produced by arc-evaporation' *J. Chem. Soc., Faraday Trans.*, **90**, 2799 (1994).

(8.8) M. S. Zwanger, F. Banhart and A. Seeger, 'Formation and decay of spherical concentric-shell carbon clusters', *J. Cryst. Growth*, **163**, 445 (1996).

(8.9) L.-C. Qin and S. Iijima, 'Onion-like graphitic particles produced from diamond', *Chem. Phys. Lett.*, **262**, 252 (1996).

(8.10) T. Füller and F. Banhart, 'In situ observation of the formation and stability of single fullerene molecules under electron irradiation', *Chem. Phys. Lett.*, **254**, 372 (1996).

(8.11) A. P. Burden and J. L. Hutchison, 'Real-time observation of fullerene generation in a modified electron microscope', *J. Cryst. Growth*, **158**, 185 (1996).

(8.12) H. W. Kroto, 'Carbon onions introduce a new flavour to fullerene studies', *Nature* **359**, 670 (1992).

(8.13) G. E. Scuseria, 'The equilibrium structures of giant fullerenes: faceted or spherical shape? An ab initio Hartree–Fock study of icosahedral C_{240} and C_{540}', *Chem. Phys. Lett.*, **243**, 193 (1995).

(8.14) Q. Ru, M. Okamoto, Y. Kondo and K. Takayanagi, 'Attraction and orientation phenomena of bucky onions formed in a transmission electron microscope', *Chem. Phys. Lett.*, **259**, 425 (1996).

(8.15) M. S. Zwanger and F. Banhart, 'The structure of concentric-shell carbon onions as determined by high-resolution electron microscopy', *Philos. Mag. B*, **72**, 149 (1995).

(8.16) C. J. Brabec, A. Maiti and J. Bernholc, 'Structural defects and the shape of large fullerenes', *Chem. Phys. Lett.*, **219**, 473 (1994).

(8.17) H. Terrones and M. Terrones, 'The transformation of polyhedral particles into graphitic onions', *J. Phys. Chem. Solids*, **58**, 1789 (1997).

(8.18) K. R. Bates and G. E. Scuseria, 'Why are buckyonions round?', *Theo. Chem. Acc.* **99**, 29 (1998).

(8.19) D. W. Thompson, *On growth and form*, Cambridge University Press, 1961.

(8.20) D. Ugarte, 'Formation mechanism of quasi-spherical carbon particles induced by electron bombardment', *Chem. Phys. Lett.*, **207**, 473 (1993).

(8.21) D. Ugarte, 'Onion-like graphitic particles', *Carbon*, **33**, 989 (1995).

(8.22) F. Banhart, T. Füller, P. Redlich and P. M. Ajayan, 'The formation, annealing and self-compression of carbon onions under electron irradiation', *Chem. Phys. Lett.*, **269**, 349 (1997).

(8.23) G. Lulli, A. Parisani and G. Mattei, 'Influence of electron-beam parameters on the radiation-induced formation of graphitic onions', *Ultramicroscopy*, **60**, 187 (1995).

(8.24) V. L. Kuznetsov, A. L. Chuvilin, Y. V. Butenko, I. Y. Mal'kov and V. M. Titov, 'Onion-like carbon from ultra-disperse diamond', *Chem. Phys. Lett.*, **222**, 343 (1994).

(8.25) V. L. Kuznetsov, A. L. Chuvilin, E. M. Moroz, V. N. Kolomiichuk, S. K. Shaikhutdinov, Y. V. Butenko and I. Y. Mal'kov, 'Effect of explosive conditions on the structure of detonation soot: ultradisperse diamond and onion carbon', *Carbon*, **32**, 873 (1994).

(8.26) T. Caboic'h, J. P. Rivière and J. Delafond, 'A new technique for fullerene onion formation', *J. Mater. Sci.*, **30**, 4787 (1995).

(8.27) T. Caboic'h, J. P. Rivière, M. Jaouen, J. Delafond and M. F. Denanot,

'Fullerene onion formation by carbon-ion implantation into copper', *Synth. Met.*, **77**, 253 (1996).

(8.28) T. Cabioc'h, M. Jaouen and J. C. Girard, 'Thin film of spherical carbon onions onto silver', *Carbon*, **36**, 499 (1998).

(8.29) Z. C. Kang and Z. L. Wang, 'Mixed-valent oxide-catalytic carbonization for synthesis of monodispersed nano sized carbon spheres', *Philos. Mag. B*, **73**, 905 (1996).

(8.30) F. Banhart and P. M. Ajayan, 'Carbon onions as nanoscopic pressure cells for diamond formation', *Nature*, **382**, 433 (1996).

(8.31) M. Zaiser and F. Banhart, 'Radiation-induced transformation of graphite to diamond', *Phys. Rev. Lett.*, **79**, 3680 (1997).

(8.32) P. Wesolowski, Y. Lyutovich, F. Banhart, H. D. Carstanjen and H. Kronmuller, 'Formation of diamond in carbon onions under MeV ion irradiation', *Appl. Phys. Lett.*, **71**, 1948 (1997).

(8.33) M. Faraday, *The chemical history of a candle*, Collier, New York, 1962.

(8.34) H. P. Palmer and C. F. Cullis, 'The formation of carbon from gases', *Chem. Phys. Carbon*, **1**, 265 (1965).

(8.35) S. J. Harris and A. M. Weiner, 'Chemical kinetics of soot particle growth', *Ann. Rev. Phys. Chem.*, **36**, 31 (1985).

(8.36) J.-B. Donnet, R. C. Bansal and M.-J. Wang, *Carbon black*, 2nd Ed., Marcel Dekker, New York, 1993.

(8.37) Q. L. Zhang, S. C. O'Brien, J. R. Heath, Y. Liu, R. F. Curl, H. W. Kroto and R. E. Smalley, 'Reactivity of large carbon clusters: spheroidal carbon shells and their possible relevance to the formation and morphology of soot', *J. Phys. Chem.*, **90**, 525 (1986).

(8.38) M. Frenklach and L. B. Ebert, 'Comment on the proposed role of spheroidal carbon clusters in soot formation', *J. Phys. Chem.*, **92**, 561 (1988).

(8.39) L. B. Ebert, 'The interrelationship of C_{60}, soot and combustion', *Carbon*, **31**, 999 (1993).

(8.40) R. Baum, 'Ideas on soot formation spark controversy', *Chem. & Eng. News*, 5 February, 1990, p. 30.

(8.41) H. Aldersey-Williams, *The most beautiful molecule*, Aurum Press, London, 1995.

(8.42) L. B. Ebert, Book review, *Carbon*, **33**, 1007 (1995).

(8.43) B. E. Warren, 'X-ray diffraction study of carbon black', *J. Chem. Phys.*, **2**, 552 (1934).

(8.44) A. E. Austin, *Proc. 3rd Conference on Carbon*, University of Buffalo, New York, 1960, p. 389.

(8.45) R. D. Heidenreich, W. M. Hess and L. L. Ban, 'A test object and criteria for high resolution electron microscopy', *J. Appl. Cryst.*, **1**, 1 (1968).

(8.46) A. B. Palotás, L. C. Rainey, C. J. Feldermann, A. F. Sarofim and J. B. Vander Sande, 'Soot morphology: An application of image analysis by high-resolution transmission electron microscopy', *Micros. Res. Tech.*, **33**, 266 (1996).

(8.47) J.-B. Donnet and E. Custodéro, 'Noir de carbone et fullérènes', *Bull. Soc. Chim. Fr.*, **131**, 115 (1994).

(8.48) J.-B. Donnet and E. Custodéro, in *Carbon black*, 2nd Ed., J.-B. Donnet, R. C. Bansal and M.-J. Wang, eds, Marcel Dekker, New York, 1993, p. 221.

(8.49) Ph. Gerhardt, S. Loffler and K. H. Homann, 'Polyhedral carbon ions in hydrocarbon flames', *Chem. Phys. Lett.*, **137**, 306 (1987).

(8.50) J. B. Howard, J. T. McKinnon, Y. Makarovsky, A. L. Lafleur and M. E. Johnson, 'Fullerenes C_{60} and C_{70} in flames', *Nature*, **352**, 139 (1991).

(8.51) P. J. F. Harris, 'Structure of non-graphitising carbons', *International Materials Reviews*, **42**, 206 (1997).

(8.52) D. Graham and W. S. Kay, 'The morphology of thermally graphitized P-33 carbon black in relation to absorbent uniformity', *J. Colloid Sci.*, **16**, 182 (1961).

(8.53) P. A. Marsh, A. Voet, T. J. Mullins and L. D. Price, 'Quantitative micrography of carbon black microstructure', *Carbon*, **9**, 797 (1971).

(8.54) H. Morrogh and W. J. Williams, 'Graphite formation in cast irons and in nickel–carbon and cobalt–carbon alloys', *J. Iron Steel Inst.*, **155**, 321 (1947).

(8.55) J. E. Harris, 'The formation of graphite in cast iron', Ph.D. thesis, University of Birmingham, 1956.

(8.56) D. D. Double and A. Hellawell, 'Cone-helix growth forms of graphite', *Acta Metall.*, **22**, 481 (1974).

(8.57) D. D. Double and A. Hellawell, 'The nucleation and growth of graphite – the modification of cast iron', *Acta Metall. Mater.*, **43**, 2435 (1995).

(8.58) B. Miao, K. Fang, W. Bian and G. Liu, 'On the microstructure of graphite spherulites in cast irons by TEM and HREM', *Acta Metall. Mater.*, **38**, 2167 (1990).

(8.59) B. Miao, D. O. Northwood, W. Bian, K. Fang and M. H. Fan, 'Structure and growth of platelets in graphite spherulites in cast iron', *J. Mater. Sci.*, **29**, 255 (1994).

(8.60) H. W. Kroto, J. P. Hare, A. Sarkar, K. Hsu, M. Terrones and R. Abeysinghe, 'New horizons in carbon chemistry and materials science', *MRS Bulletin*, **19(11)**, 51 (1994).

(8.61) J. P. Hare, K. Hsu, M. Terrones, A. Sarkar, S. G. Firth, A. Lappas, R. Abeysinghe, H. W. Kroto, K. Prassides, R. Taylor and D. R. M. Walton, 'Fullerene-based materials science at Sussex', *Molecular Materials*, **7**, 17 (1996).

(8.62) M. B. Haanstra, W. F. Knippenberg and G. Verspui, *Proc 5th European Conference on Electron Microscopy*, Manchester, Institute of Physics, 1972, p. 214.

(8.63) H. Marsh (ed.), *Introduction to carbon science*, Butterworth, London, 1989.

(8.64) J. D. Brooks and G. H. Taylor, 'The formation of some graphitizing carbons,' *Chem. Phys. Carbon*, **4**, 243 (1968).

9

Future directions

What's past is prologue.
William Shakespeare, *The Tempest*

Phenomenal progress has been made in the science of carbon nanotubes since the publication of Iijima's landmark paper in 1991. We now know how to prepare nanotubes in bulk and how to purify them; how to prepare single-walled as well as multiwalled tubes; how to produce layers of aligned tubes on substrates; and how to remove the end-caps and introduce foreign materials inside. Theoreticians have predicted that nanotubes should have extraordinary electronic properties, and many of these predictions have been confirmed experimentally. The mechanical properties of nanotubes have proved to be equally exceptional, and nanotube-containing composites are now exciting great interest. Nanotube research is burgeoning, with papers currently appearing at the rate of more than one per day. Nevertheless, it is quite conceivable that we have only scratched the surface of what may be possible in the new carbon science.

This final chapter considers some possible directions in which the subject might develop. Firstly, the development of a nanotube-based organic chemistry is discussed. Some tentative steps have already been taken in this direction, with the demonstration that nanotubes can be functionalised, and that functional groups can be removed, leaving the original nanotube structure intact. Secondly, some possible new, all-carbon, molecules with nanotube-based structures are described. This area is at present purely theoretical, since the techniques for synthesising these all-carbon structures have not yet been developed. Finally, the idea that nanotube-based structures might provide components for the nanomachines of the distant future is briefly discussed.

9.1 Towards a carbon nanotube chemistry

The bulk synthesis of C_{60} in 1991 was rapidly followed by the development of new branches of organic and organometallic chemistry based on the fullerenes. The chemistry of nanotubes has been slower to evolve, mainly because the tubes have much less well-defined structures than do fullerenes. Nevertheless, several groups are now working on the chemical modification of nanotubes, and some promising results are being reported.

One very basic aspect of nanotube chemistry was established in the early days of the subject: they are most reactive at the tips. This was demonstrated during experiments by Iijima's group in Japan and by Tsang, Green and colleagues in Oxford aimed at opening and filling the tubes (see Chapter 5). It was found that treatment of nanotubes with mild oxidising agents resulted in corrosion or complete removal of the caps, but that the cylindrical parts of the tubes were left untouched. In the case of treatments with acid, the selectivity could be extremely high. This enhanced chemical reactivity is a consequence of the presence of pentagonal rings and other non-hexagonal rings in the caps.

Experiments on opening multiwalled tubes with acids and other reagents also showed that such treatments could result in the formation of surface carboxylate and other groups on the nanotube surfaces. More recently, attempts have been made to functionalise single-walled nanotubes. For example, in late 1998 a group from the University of Kentucky described a method for dissolving single-walled tubes in organic solutions by derivatisation with thionylchloride and octadecylamine (9.1). This approach opens the way for solution-phase chemistry to be carried out on single-walled nanotubes. At about the same time, workers from Rice University described experiments on the fluorination and defluorination of single-walled tubes (9.2). Measurements using electron microscopy and Raman spectroscopy indicated that in many cases the tubes were restored to their original structures following defluorination, demonstrating that reversible chemistry on nanotubes is a possibility.

Another milestone was the demonstration by the Rice group that single-walled tubes can be broken down into short lengths ('fullerene pipes') and that these open tubes could be sorted into different length fractions (Chapter 2). Although the resulting tubes would be expected to have a range of different atomic structures, it is conceivable that methods could be devised for sifting them into structural types (perhaps by exploiting their different electronic properties?). If so, then the way would be open to a true nanotube-based organic chemistry.

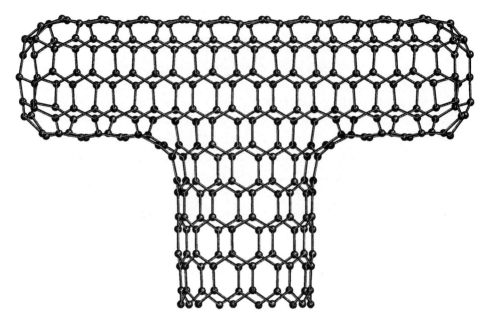

Fig. 9.1. Simulation of T-junction between (5,5) and (10,0) tubes (9.6).

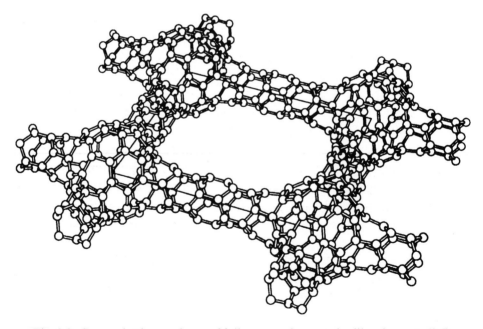

Fig. 9.2. Square lattice made up of fullerene and nanotube-like elements (9.5).

Fig. 9.3. Molecular bearing based on two nanotube-like cylinders (9.9).

9.2 New all-carbon structures

As noted at the beginning of this book, imaginary all-carbon crystals and molecules have been discussed by theoreticians for decades. Since the discovery of C_{60}, many new all-carbon frameworks with fullerene-related structures have been postulated, notably the so-called schwarzites (9.3). A number of authors have also considered theoretical structures based on nanotubes (e.g. 9.4–9.6). These structures usually involve junctions containing non-hexagonal rings. For example, Madhu Menon of the University of Kentucky and Deepak Srivastava of the NASA Ames Research Center have considered the properties of nanotube 'T-junctions' (9.6). One such junction, between a (5,5) armchair tube and a (10,0) zig-zag tube is shown in Fig. 9.1. This junction contains six heptagons, arranged in two groups of three. Different arrangements of non-hexagonal rings can result in 'Y-junctions'. Junctions rather similar to some of these (albeit with far less perfect structures) have been observed experimentally (p. 95).

More extensive nanotube-related frameworks were discussed by Noriaki Hamada (9.5), who considered 2D and 3D structures based on very narrow tubes. A part of one of his 2D lattices is shown in Fig. 9.2. Structures of this kind could provide the basis for nanoscale electronic devices – but first we need to find a way to synthesise them.

9.3 Nanotubes in nanotechnology

Nanotechnology has been the subject of much scientific and pseudo-scientific speculation in recent years (9.7, 9.8). This speculation has been partly inspired by the development of instruments such as the scanning tunnelling microscope, which seem to have the capacity to construct nanomachines atom-by-atom. These machines, it is suggested, could revolutionise manufacturing industry, medicine and almost every other aspect of our lives. When considering which materials might be used in the construction of such devices, many nanotechnologists have favoured diamond, and have shown how components such as bearings could be made from sp^3-bonded carbon. However, sp^2 carbon would seem to have equal potential in this area, and several groups have considered the idea that nanotubes might be used in nanomechanical devices.

The simple carbon nanotube 'bearing' shown in Fig. 9.3 was discussed by Robert Tuzun and colleagues from Oak Ridge National Laboratory (9.9). This bearing consists of a very narrow inner cylinder (shaft), and a shorter outer cylinder (sleeve), separated by a distance slightly larger than the graphite interplanar spacing. Simulations suggested that such an assembly could function reasonably well as a bearing, but that performance might be limited by vibrational effects.

Workers from the Naval Research Laboratory in Washington have considered more complex nanocomponents based on fullerene and nanotube structures (9.10). One of these assemblies, a six-tooth sprocket with a shaft based on a (6,6) armchair nanotube, is shown in Fig. 9.4. Calculations suggested that these structures should be reasonably stable, and modelling indicated that the gears could be turned against each other at high angular velocities without damage.

An alternative type of nanotube gear was discussed by a group from NASA (9.11). In this case, the gears were produced by functionalising tubes with benzyne 'teeth', as shown in Fig. 9.5. Such structures might be more amenable to synthesis than the assemblies of the NRL group, and once again simulations indicated that they could operate at high speed without major deformations. Needless to say, however, the idea that such nanomachines could ever become a reality is highly controversial.

9.4 Final thoughts

If there is one thing which has characterised fullerene and nanotube science, it is serendipity. The discovery of buckminsterfullerene itself was a wonderful

(a)

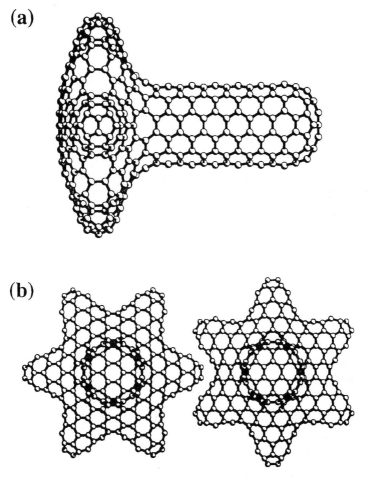

(b)

Fig. 9.4. (a), (b) Side view and top view of six-tooth gear with nanotube shaft (9.10).

accident, and nanotubes were an unanticipated byproduct of the bulk synthesis of C_{60}. Other fruits of serendipity include single-walled nanotubes, inorganic fullerenes and carbon onions. Each of these developments has opened up a new area of enquiry, and frequently led to further unexpected discoveries. Ultimately, therefore, attempting to predict future directions in nanotube science is probably futile. We are still in the early days of the field, and many surprises undoubtedly lie ahead.

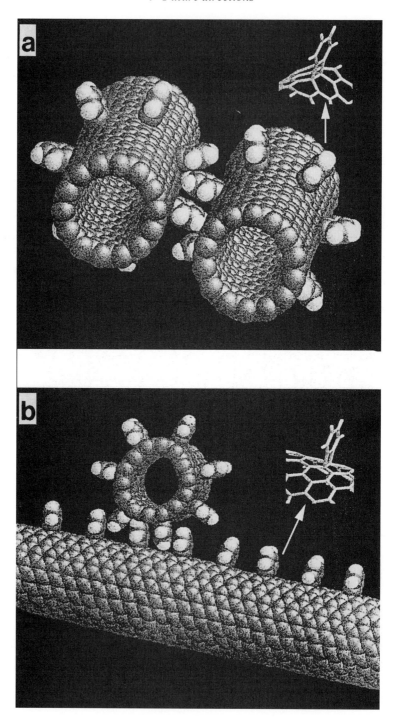

Fig. 9.5. (a) Carbon nanotube-based gears with benzyne teeth, (b) gear and shaft system (9.11).

References

(9.1) J. Chen, M. A. Hamon, H. Hu, Y. S. Chen, A. M. Rao, P. C. Eklund and R. C. Haddon, 'Solution properties of single-walled carbon nanotubes', *Science*, **282**, 95 (1998).

(9.2) E. T. Mickelson, C. B. Huffman, A. G. Rinzler, R. E. Smalley, R. H. Hauge and J. L. Margrave, 'Fluorination of single-wall carbon nanotubes', *Chem. Phys. Lett.*, **296**, 188 (1998).

(9.3) H. Terrones, M. Terrones and W. K. Hsu, 'Beyond C_{60} – graphite structures for the future', *Chem. Soc. Rev.*, **24**, 341 (1995).

(9.4) L. A. Chernozatonskii, 'Carbon nanotube connectors and planar jungle gyms', *Phys. Lett. A*, **172**, 173 (1992).

(9.5) N. Hamada, 'Electronic band-structure of carbon nanotubes: toward the three-dimensional system', *Mater. Sci. Eng. B*, **19**, 181 (1993).

(9.6) M. Menon and D. Srivastava, 'Carbon nanotube "T junctions": nanoscale metal–semiconductor–metal contact devices', *Phys. Rev. Lett.*, **79**, 4453 (1997).

(9.7) K. E. Drexler, *Engines of creation – The coming era of nanotechnology*, Doubleday, New York, 1986.

(9.8) E. Regis, *Nano! – Remaking the world atom by atom*, Bantam, London, 1995.

(9.9) R. E. Tuzun, D. W. Noid and B. G. Sumpter, 'The dynamics of molecular bearings', *Nanotechnology*, **6**, 64 (1995).

(9.10) D. H. Robertson, B. I. Dunlap, D. W. Brenner, J. W. Mintmire and C. T. White, in *Novel forms of carbon II*, Eds. C. L. Renschler *et al.*, *Mat. Res. Soc. Symp. Proc.*, Vol. 349, 1994, p. 283.

(9.11) J. Han, A. Globus, R. Jaffe and G. Deardorff, 'Molecular dynamics simulation of carbon nanotube based gears', *Nanotechnology*, **8**, 95 (1997).

Name index

Subject index